Establishing a Performance Index for Construction Project Managers

Assessment of professional competence for project managers and the measure of project success is well-trodden ground in the research and professional project management literature. Whilst standards and certifications like PMBOK and the IPMA competence baseline have been developed as a guide for the development of project managers' competence, the manifestation of these competencies into good performance is neither guaranteed nor always easily ascertainable.

This book presents a brand new, comprehensive, and reliable quantitative tool to assess the performance of a construction project manager. Though the performance of a construction project manager may be judged on time and cost criteria of a project, there is still no one conclusive evaluation tool based on the varied criteria or competencies that are usually ascribed to them.

This book develops a performance index for construction project professionals which can be indicative of their performance measured over varied attributes over the lifetime of their professional development. This index has the potential to provide all project stakeholders with better control over selecting appropriate resources for managing projects and drive the project professional from within towards improving his/her credentials with every project.

This book can be used by aspiring and practising project managers for measuring their own performance and assessing their relative strengths and weaknesses. Organizations can use the tool as a benchmark to select the best of their human resources for their projects, and training institutions can use the tool to set a baseline, highlight areas for intervention, and indicate the readiness of trainees to face real world projects.

Prof. Dr. Virendra Kumar Paul is faculty at the Department of Building Engineering and Management at the School of Planning and Architecture, New Delhi. Associated with the department since 1985, his areas of interest in research and academia include construction project management and fire safety in buildings. He is a well-published author in renowned high-impact journals such as Springer International, Taylor & Francis, UGC Care, and other peer-reviewed journals.

Sushil Kumar Solanki is faculty at the Department of Building Engineering and Management, School of Planning and Architecture, New Delhi. He has lectured extensively on construction management in India and internationally, and has several research papers in various peer-reviewed journals and conferences to his credit. His primary academic and research interests span from construction project management and building services to building rehabilitation.

Dr. Abhijit Rastogi is faculty at the Department of Building Engineering and Management, School of Planning and Architecture, New Delhi. Having been a part of the academia for more than 8 years, Dr. Abhijit is extensively involved in research and teaching. His areas of academic research include construction management, project scheduling, building materials, and skill development.

Parnika Singh Yadav is research scholar at the Department of Building Engineering and Management, School of Planning and Architecture, New Delhi. Her areas of interest include construction project management and risk assessment. Her current area of research is project complexity management. She has also published papers in peer-reviewed high-impact journals.

Establishing a Performance Index for Construction Project Managers

Virendra Kumar Paul,
Sushil Kumar Solanki,
Abhijit Rastogi and
Parnika Singh Yadav

LONDON AND NEW YORK

Cover image: © Bim/Getty Images

First published 2023
by Routledge
4 Park Square, Milton Park, Abingdon, Oxon OX14 4RN

and by Routledge
605 Third Avenue, New York, NY 10158

Routledge is an imprint of the Taylor & Francis Group, an informa business

© 2023 Virendra Kumar Paul, Sushil Kumar Solanki,
Abhijit Rastogi & Parnika Singh Yadav

The right of Virendra Kumar Paul, Sushil Kumar Solanki,
Abhijit Rastogi & Parnika Singh Yadav to be identified as authors
of this work has been asserted in accordance with sections 77 and
78 of the Copyright, Designs and Patents Act 1988.

All rights reserved. No part of this book may be reprinted or
reproduced or utilised in any form or by any electronic, mechanical,
or other means, now known or hereafter invented, including
photocopying and recording, or in any information storage or
retrieval system, without permission in writing from the publishers.

Trademark notice: Product or corporate names may be trademarks
or registered trademarks, and are used only for identification and
explanation without intent to infringe.

British Library Cataloguing-in-Publication Data
A catalogue record for this book is available from the British Library

Library of Congress Cataloging-in-Publication Data
Names: Paul, Virenda Kumar, author. | Solanki, Sushil Kumar, author. |
Rastogi, Abhijit, author. | Yadav, Parnika Singh, author.
Title: Establishing a performance index for construction project
managers / Virenda Kumar Paul, Sushil Kumar Solanki, Abhijit
Rastogi, Parnika Singh Yadav.
Description: Abingdon, Oxon ; New York, NY : Routledge, [2023] |
Includes bibliographical references and index.
Identifiers: LCCN 2022042250 (print) | LCCN 2022042251 (ebook) |
ISBN 9781032345529 (pbk) | ISBN 9781032345574 (hbk) |
ISBN 9781003322771 (ebk)
Subjects: LCSH: Construction workers—Job descriptions. | Project
managers—Job descriptions. | Project managers—Rating of. |
Building—Superintendence. | Performance standards. |
Core competencies.
Classification: LCC TH159 .P38 2023 (print) | LCC TH159 (ebook) |
DDC 690.23—dc23/eng/20221115
LC record available at https://lccn.loc.gov/2022042250
LC ebook record available at https://lccn.loc.gov/2022042251

ISBN: 978-1-032-34557-4 (hbk)
ISBN: 978-1-032-34552-9 (pbk)
ISBN: 978-1-003-32277-1 (ebk)

DOI: 10.1201/9781003322771

Typeset in Times New Roman
by codeMantra

Contents

List of figures	ix
List of tables	xi
Preface	xiii
Acknowledgements	xvii
List of abbreviations	xix

1 Construction project manager: role and practices   1

  1.1 Introduction 1
    1.1.1 Project management 1
    1.1.2 Waterfall project management 2
    1.1.3 Agile project management 2
    1.1.4 Lean project management 2
  1.2 Construction project management and its peculiarities 2
    1.2.1 Role of construction project manager vis-à-vis other stakeholders 3
  1.3 General challenges of CPM 3
  1.4 Challenges of CPM in Indian construction industry 4
  1.5 Process of construction project management followed in India 5
  1.6 Validation of need for construction management performance 5
  1.7 Prevailing practice of project management in construction 7
  1.8 Inferences 14

2 Performance assessment of construction project managers: practices and challenges   16

  2.1 Introduction 16
  2.2 Limitations of existing standards 16
  2.3 Practice of CPM in private sector in India: who, where from, and what he/she does 18
  2.4 Practice of CPM in public sector in India: who, where, from, and what he/she does 19

vi   *Contents*

    2.5 Existing body of knowledge for competence development 19
        2.5.1 Project manager competency development framework by PMI 20
        2.5.2 Association of project management competency framework 25
        2.5.3 Australian Institute of Project Management (AIPM) professional competency management standards 25
        2.5.4 Complex project manager competency standards 25
        2.5.5 360-degree perspective from stakeholders 28
    2.6 Performance versus competence 30
    2.7 Qualification 30
    2.8 Skill 32
    2.9 Processes 33
    2.10 Competence 33
    2.11 Barriers in performance 33
    2.12 Way ahead through performance assessment 35
    2.13 Role, responsibility, and accountability 36
    2.14 Value addition through construction project managers 40
    2.15 Inferences 41

3 Performance index for construction project managers   46
    3.1 Introduction 46
    3.2 Concept of value drivers performance index (VDPI) 46
    3.3 Deriving variables for VDPI 47
    3.4 Quantification process 50
        3.4.1 Performance indicators and their determinants 52
    3.5 Evaluation process of VDPI 55
        3.5.1 Multicollinearity of performance indicators 56
        3.5.2 Relationship of variables across project lifecycle 62
        3.5.3 Limitations 69
    3.6 Application of VDPI 70
        3.6.1 Process flow for VDPI hardware interface 75
    3.7 Inferences 77

4 Value-driven performance assessment of construction project managers   79
    4.1 Introduction 79
    4.2 Performance indicators 79
        4.2.1 Need for performance indicators 79
        4.2.2 Meaning of performance indicators 80
        4.2.3 Traditional project performance indicators 82
        4.2.4 Use of performance indicators 83

Contents vii

4.2.5 Time management performance indicators and determinants 87

4.2.6 Cost management performance indicators and determinants 91

4.2.7 Scope management performance indicators and determinants 95

4.2.8 Contract management performance indicators and determinants 102

4.2.9 Design management performance indicators and determinants 107

4.3 Inferences 112

5 Threshold performance level for time management performance 117

5.1 Introduction 117

5.2 Criterion for defining five levels of threshold performance 117

5.3 Classification of time performance threshold levels 118

5.3.1 Planning work coordination (W_{11}) 118

5.3.2 Effective schedule control (W_{12}) 120

5.3.3 Risk forecasting (W_{13}) 120

5.3.4 Effective resource planning (W_{14}) 121

5.3.5 Controlling delays (W_{15}) 122

5.4 Inferences 127

6 Threshold performance level for cost management performance 128

6.1 Introduction 128

6.2 Classification of cost performance threshold levels 128

6.2.1 Effective cash flow management (W_{21}) 128

6.2.2 Controlling budget variance (D_{22}) 130

6.2.3 Managing risk contingencies (D_{23}) 131

6.2.4 Controlling cost overruns (D_{24}) 132

6.3 Inferences 135

7 Threshold performance level for scope management performance 136

7.1 Introduction 136

7.2 Classification of scope performance threshold levels 136

7.2.1 Coordinated scope planning (W_{31}) 136

7.2.2 Effective stakeholder involvement (W_{32}) 138

7.2.3 Monitoring project deliverables (W_{33}) 139

7.2.4 Controlling scope creep (W_{34}) 140

7.3 Inferences 144

viii *Contents*

8 Threshold performance level for contract management performance   145

   8.1 Introduction 145

   8.2 Classification of contract management performance
      threshold levels 145

      8.2.1 Risk-sensitive procurement planning (W_{41}) 145
      8.2.2 Planning contractual obligations (W_{42}) 147
      8.2.3 Managing contractual obligations (W_{43}) 148
      8.2.4 Effective claim management (W_{44}) 148
      8.2.5 Planning contract closeout (W_{45}) 149

   8.3 Inferences 154

9 Threshold performance level for design management performance   155

   9.1 Introduction 155

   9.2 Classification of design performance threshold levels 155

      9.2.1 Establishing stakeholder engagement
            processes (D_{51}) 155
      9.2.2 Establishing need centric design process (D_{52}) 157
      9.2.3 Establishing decision-making hierarchy (D_{53}) 158
      9.2.4 Resolving conflicting interests (D_{54}) 159
      9.2.5 Effective planning for scope creep (D_{55}) 160
      9.2.6 Resolving time-cost impact (D_{56}) 161

   9.3 Inferences 165

10 Way forward through complexity linkage   166

   10.1 Introduction 166

   10.2 Conclusion and way forward through complexity
      linked with VDPI 166

      10.2.1 Concept of project complexity indicator (PCI) 170
      10.2.2 Convergence of VDPI and PCI 172
      10.2.3 PCI calculation methodology 173
      10.2.4 Process of deriving variables for PCI equation 177
      10.2.5 Level of complexity / PCI range 179

   10.3 Inferences 179

   Index   181

Figures

1	Typical organizational structure of design bid build model	8
2	Practice of project managers in organization hierarchy	8
3	Level of CPM's competence needed based on the emerging complexities out of stakeholder's interaction	10
4	Extent of client interaction and level of competence of CPM	11
5	General organizational hierarchy matrix representing stakeholder interaction	12
6	Developer-based organizational hierarchy model	13
7	Project manager competency development model	21
8	Project manager performance assessment criterion example in case of construction project	23
9	Competence development process	24
10	Performance and competence-based diagnosis	30
11	Value driver performance index (VDPI) concept	47
12	VDPI process methodology flowchart	57
13	Correlation between the variables	60
14	Relationship of variables across project lifecycle	63
15	Correlations between the independent variables are not accounted for in existing model	69
16	Hardware device processing phase process for IU	71
17	Hardware device processing Phase Plan-Project Host (PH)	72
18	Hardware device processing Phase Plan-Organization Unit (OU)	73
19	Hardware device processing Phase Plan-Multiportfolio Organization Unit (Orgn Unit)	74
20	Hardware device support for operationalizing VDPI	76
21	Dimensions of project success	85
22	How performance indicators support management actions	86
23	Performance indicators of construction projects	87
24	Determinants of time management performance and PMBOK process group	92
25	Determinants of cost management performance and PMBOK process groups	96

x *Figures*

26	Determinants of scope management performance and PMBOK process groups	102
27	Contract management within a construction project lifecycle	103
28	Determinants of contract management performance and PMBOK process groups	107
29	A graphical representation of comparison VDPI score obtained by a CPM "X" through self-assessment, benchmarking, and project complexity requirement	169
30	Drivers of project complexity	170
31	Project success as a function of VDPI and project complexity	172
32	Mapping concept of VDPI and PCI levels	174
33	Taxonomy of project complexity	175
34	Matrix for determining C_{11}, C_{12},...,C_{1N} sub-determinant values	178

Tables

1	Range indicators for each management unit	26
2	Performance versus competence	31
3	Responsibilities of a project manager as per CIOB	39
4	A summarized view of the indicators and their determinants is produced in the form of a matrix	54
5	Relationship of time performance variables across project lifecycle	64
6	Relationship of cost performance variables across project lifecycle	65
7	Relationship of scope performance variables across project lifecycle	66
8	Relationship of contract performance variables across project lifecycle	67
9	Relationship of design performance variables across project lifecycle	68
10	Hardware device interface and their functions	75
11	Concept of use of leading and lagging indicators	81
12	KPIs considered in past studies for evaluating the performance of construction projects	84
13	Tools and techniques of project performance related to time as per PMBOK@ Guide	89
14	Mapping of time management processes and determinants	89
15	Determinants of time performance	91
16	Indicators of project performance related to cost as per PMBOK@ Guide	93
17	Mapping of cost management processes and determinants	94
18	Determinants of cost performance	96
19	Tools and techniques of project performance related to scope as per PMBOK@ Guide	97
20	Mapping of scope management processes and determinants	99
21	Determinants of scope performance	102
22	Mapping of contract management processes and determinants	104

xii *Tables*

23	Determinants of contract performance	107
24	Stages of RIBA plan of work 2020	108
25	Initial design evaluation criteria	109
26	Determinants of design performance	112
27	Performance levels	118
28	Threshold performance levels for time management performance indicator determinants	124
29	Threshold performance levels for cost management performance indicator determinants	133
30	Threshold performance levels for scope management performance indicator determinants	142
31	Threshold performance level of contract performance indicator	150
32	Threshold performance levels of design performance indicator	162
33	Questionnaire sample question	178

Preface

The exponential growth of the construction industry globally is playing a momentous role in propelling the development of the nations. With the projected contribution to GDP to be at around 13% by 2050, the construction industry became the second-largest FDI equity recipient sector for India in 2020–2022. Additionally, the real estate sector in the Indian market is expected to reach the market size of 1USD Tn by 2030. With such advancements in the financial and technological systems of construction projects, processes and implementation mechanisms are turning out to be dynamic and uncertain, thereby giving rise to the need of project managers with relevant skill sets enabled to accomplish such construction projects. Further, the performance of an organization is driven by the successful delivery of construction projects, which in turn is significantly dependent on construction project manager's performance. With increasing competition and onset of new organizations, it is important to ameliorate the performance of a construction project manager to achieve better performance.

It is, therefore, crucial to maintain and enhance the performance of projects and organizations in order to maintain competitiveness within the construction industry in the long run. The evaluation of performance of each employee/manager handling projects is important for the determination of the individual as well as team shortcomings and areas for improvement. Moreover, the evaluation of performance also helps the individuals to self-assess their performance as well as managers to evaluate the performance of other individuals based on the projects handled by them. This approach is also useful during the appraisal process of individuals working in the organization. Hence, organizations need to identify and evaluate the key indicators to measure the performance of the organization as well as individual employees at different stages, thereby providing a scope of improvement within the organization in specific areas.

Several standards related to competency assessment and development have been formulated by international organizations such as the Project Management Institute (PMI) and the Australian Institute of Project

xiv *Preface*

Management (AIPM). In the academia and research sector, tools like Leadership Dimensions Questionnaire (LDQ), the McBer Competency Framework, the Mayer-Salovey-Caruso Emotional Intelligence Test, the General Mental Ability (GMA), the Inwald Personality Inventory (IPI), the Multifactor Leadership Questionnaire (MLQ), and the Myers-Briggs Type Indicator (MBTI) have been devised to assess project managers' competencies. However, it has been observed that these standards and tools do not provide a quantitative index, thereby proving to be inadequate in achieving a more accomplished set, in order to fulfil the project objectives.

To accomplish the same, this book establishes the concept of value driver performance index (VDPI) for assessing the performance of a construction project manager at individual and organizational levels. The process of deriving variables for defining VDPI has been covered in detail in this book. It comprises of the terminologies used for defining VDPI, process of deriving variables for developing the performance index, quantification process, and the evaluation process for determining the VDPI value. The notion of multicollinearity has been briefly discussed in the developed VDPI equation indicating the interrelationship of the derived variables across the different stages of project lifecycle.

The book also discusses about the strategies for enhancing the VDPI and establishes a synopsis of existing body of knowledge for competence development. Further, benchmarking as a tool for continuous improvement is discussed, to assess how the benchmarking approach can be utilized in order to assess and improve the construction project management performance. The book also presents examples related to the VDPI calculation process and how it can be used as a software-based application for assessing the project manager's performance. The software-based application can be used as a method for evaluating the performance of an organization, which involves entering input data by an employee into a software module of the system via an operating device; analysing the input data of the employee by the software module of the system; generating performance indicators and corresponding determinants by the software module of the system; displaying performance indicators and corresponding determinants through the graphical user interface of the software module on a display of the operating device; scoring the determinants of each performance indicator by the employee through the operating device based on their performance with respect to each determinant of the performance indicator; evaluating numerical value of each performance indicator by the software module from weighed summation of the scores accorded to each determinant by the employee; displaying the evaluated value of the performance indicators through graphical user interface of software module on the display of the operating device; evaluating the value of performance index of the employee by the software module from weighed summation of scores achieved under each performance indicators; displaying the evaluated value

Preface xv

of the performance index through graphical user interface of software module on the display of the operating device; and storing the value of performance index evaluated into a server of the system.

The authors hope that this book would be perceived as a leading document in the field of construction project management performance assessment and will set a precedent as being technically accurate yet coherent and unified.

Acknowledgements

We wish to express our profound gratitude to the School of Planning and Architecture, New Delhi for giving us the opportunity to conduct this research. As the book builds on the concept of project management and performance assessment, the authors wish to record our acknowledgement to the previous contributors in the field. The idea of writing the book grew out of the belief that it's the people of the organization who act as building blocks and define the success of an organization which needs to be delved deeper into for the ease of its application.

We would like to express our sincere gratitude and deepest regards to our colleagues in the Department of Building Engineering and Management – Mr. Salman Khursheed, Mr. Luke Judson, Mr. Amit Moza, Ms. Sumedha Dua, Ms. Devika Nayal, and Mr. Kuldeep Kumar – for their immense help, guidance, and their valuable and critical inputs during our periodic discussions.

Our appreciation also extends to the fraternity of construction industry professionals who participated in the making of this book through their valuable contributions during data collection and analysis stage.

Abbreviations

A

Actual Cost of the Work Performed	(ACWP)
Architecture, Engineering and Construction	(AEC)
Artificial Intelligence	(AI)
Association of Project Management	(APM)
Australian Institute of Project Management	(AIPM)

B

Billing Performance Index	(BPI)
Building Information Modelling	(BIM)

C

Central Public Works Department	(CPWD)
Certified Associate in Project Management	(CAPM)
Certified Construction Manager	(CCM)
Cost Performance Index	(CPI)
Cost Variance	(CV)
Cultural Complexity	(CC)

E

Earned Revenue of the Work Performed	(ERWP)
Engineering, Procurement, Construction	(EPC)
Environmental complexity	(EC)
Extension of Time	(EOT)

F

Foreign Direct Investment	(FDI)

xx *Abbreviations*

G

General Mental Ability (GMA)
Gross Domestic Product (GDP)

I

Information complexity (IC)
Input unit (IU)
International Organization for Standardization (ISO)
Inwald Personality Inventory (IPI)

K

Key Performance Indicators (KPI's)

L

Leadership Dimensions Questionnaire (LDQ)
Legal complexity (LC)

M

Multifactor Leadership Questionnaire (MLQ)
Multiportfolio CPM (Orgn Unit)
Myers-Briggs Type Indicator (MBTI)

O

Organization unit (OU)
Organizational Complexity (OC)

P

Performance Indicators (PIs)
Portfolio Management Professional (PfMP)
Program Management Professional (PgMP)
Project Complexity (PC)
Project complexity success indicator (PCSI)
Project host (PH)
Project Management Certification (PMP)
Project Management Institute (PMI)
Project Management Professional (PMP)
Public Sector Units (PSU)

Abbreviations xxi

R

Request formation information	(RFI)
Return on investment	(ROI)
Royal Institute of Charted Surveyors	(RICS)

S

Schedule performance index	(SPI)
Schedule variance	(SV)

T

Task complexity	(TC)
Technological complexity	(TEC)
The American Concrete Institute	(ACI)
Trillion	(Tn)

U

Uncertainty complexity	(UC)
US Dollar	(USD)

V

Value drivers performance index	(VDPI)

W

Work breakdown structure	(WBS)

1 Construction project manager
Role and practices

1.1 Introduction

This chapter establishes the need for assessment of the performance of a construction project manager (CPM). Project management approaches and processes used in the construction industry are discussed in brief, in addition to a detailed description of the relevance of construction project management in contemporary construction. The chapter brings forth a comprehensive definition of a CPM, along with the identification of challenges faced by the CPM. The issues related to the prevailing practice of construction management in India are also included in this chapter. Further, the need for performance assessment of the CPM is extruded using the available studies, thereby leading to a redefining of the CPM, based on organizational hierarchy and project development models adopted.

1.1.1 Project management

Projects constitute a number of interrelated elements, have fixed timelines, have a definite budget, consume committed resources, and are subject to varying project environments in which they are being implemented. The term project management is defined in various literatures and generally concurs on a definition as *the process which involves planning and organizing of the resource to deliver or achieve a set of objectives required for successful handover of projects.* Additionally, it can be defined as a set of activities that are adopted to ensure the successful delivery of projects. It may involve a one-time project or an ongoing activity. The application of project management processes lies within the domain of different fields like engineering, construction, healthcare, facility management, and information technology which involve a complex set of deliverables to be achieved with specialized expertise within a stipulated budget and time period. By itself, project management has no existence. Typically, a project involves a start-to-finish plan which provides an outline for the kick-off of the project, management approach, resources allocation strategy, and project handover. The application of the project management process helps in keeping the project on track

DOI: 10.1201/9781003322771-1

2 Construction project manager: role and practices

while ensuring the timely delivery of the targeted deliverables. Based on the typology of projects and industry-specific needs, different types of project management processes have been developed. However, PMBoK, PRINCE, etc. provide a common set of industry practice frameworks, adapted for different domains.

1.1.2 Waterfall project management

A traditional approach of project management, waterfall project management works on the approach of linear assumption caveat that one activity can start after the accomplishment of one task. The assumption of linearity in the complete flow of sequence of tasks and the progress of tasks is in one direction only, similar to the flow of a river.

1.1.3 Agile project management

The process of agile project management approach does not follow the sequential linear process flow for achieving work progress; instead, it assumes that the tasks can be executed parallel to each other. This ensures the identification of errors and issues during the process of parallel execution without requiring the complete process to be restarted. The agile management approach finds its major application in the information technology industry. It can be described as an iterative process focusing on continuous improvement of the overall process and deliverables.

1.1.4 Lean project management

The concept of lean project management was first gleaned by Japanese manufacturing practices. The inkling behind the concept of lean management is to enhance the efficiency in order to reduce the wastage of time and resources.

Project management processes continue to evolve with the emerging technologies and growing project-specific requirements, thereby leading to advanced processes.

1.2 Construction project management and its peculiarities

Considering the complex nature of construction projects, the process of construction management is crucial for their successful completion with CPM playing the key role. The peculiarities found in construction projects can significantly affect projects success and failure. The management of different units of construction projects requires an established management process which is termed as construction project management.

According to CMAA, construction management can be understood as a professional service that is aimed at providing the project owner(s) with effective management of the cost, scope, schedule, quality, safety, and function (CMAA, 2022).

Construction management acts as the backbone of any project. Construction projects vary with respect to their typology, size, timeline, budget, and environment. Every project is unique and is governed by its own characteristics based on the project requirements, project delivery methods, etc. As construction projects involve a large number of variables specific to projects, the type of technology, skillset needed, and administration requirements vary accordingly and require appropriate structured, disciplined management of multiple project variables to create value.

Considering the ambiguity involved in construction projects and their emerging nature, the application of appropriate project management approach is extremely critical to ensure effective management and completion of the project. Hence, construction management can be considered to be a professional service.

1.2.1 Role of construction project manager vis-à-vis other stakeholders

The role of a CPM entails converging the interest of all stakeholders involved in the project towards achieving project's success and ensuring that all aspects related to the scope of work of all engaged stakeholders are well understood and are aware about their contractual obligations. The role of a CPM is based on the necessary principle of project management governed by the iron triangle of time, cost, and scope. It also includes other parameters of quality, design, safety, and contract. CPM provides clear lines of accountability to all project stakeholders in addition to his/her responsibility for analyzing the project in terms of deliverables, timelines, budget, and the associated risks. As a project manager, one has the ability to identify emerging risks and provide valuable mitigation strategies. The involvement of a CPM in the early phase of the project ensures the formulation of appropriate project management plan which takes into account the project teams responsibility as a whole and their deliverables.

1.3 General challenges of CPM

According to the 2021 Talent Gap report by Project Management Institute (PMI), it has been estimated that over 61 million project management positions in construction and manufacturing will be required by 2030, which indicates a 13% increase since 2019. Hence, a major challenge faced by the construction industry in terms of project management is the lack of technically well-equipped project managers. The first order of any business considering new advancements is to embrace the upcoming technologies like artificial intelligence, drones, deep machine learning, robotics, and mobile applications for achieving efficient project deliveries.

4 *Construction project manager: role and practices*

The uniqueness of the project is also one of the challenges in the traditional methodology of project management as every project needs to be treated independently based on the project characteristics. In construction industry, it has become extremely challenging to manage projects as all project team members and stakeholders must comprehend the concepts of adaptability and flexibility.

With the onset of new technologies, it is foremost to adopt them for the smooth functioning of a project. Construction companies are already experimenting with robotics, drones, artificial intelligence, mobile apps, and cloud storage to some level. In addition to these, other technologies such as Internet of Things, 5-G building information modelling, and high-definition surveying and geolocation are anticipated to have tremendous potential. More technologically progressive construction teams and a commitment to upskilling employees will be necessary to adopt and implement these technologies, which is a major challenge for the construction industry.

1.4 Challenges of CPM in Indian construction industry

Construction industry plays a very important role in propelling the development of the nation and its contribution to the Gross Domestic Product (GDP) of the country is expected to be around 13% by 2050. It is the second largest Foreign Direct Investment (FDI) equity recipient sector for India in 2020–2022, the real estate sector in the Indian market by 2030 is expected to reach the market size of 1 US Dollar (USD) Trillion (Tn). The construction sector is also the second largest employer in the country as well (Construction | Make in India, 2022). With a potential for tremendous growth, the construction industry acts as one of the parameters for defining the nation's economic progress. However, it needs to be managed in an enhanced formalized manner since the industry faces major challenges of cost and time overruns in projects. Construction industry is labour intensive and unstructured, and the dearth of accomplished project managers and skilled labour poses major challenges to the successful and timely completion of the project. In the Indian context, the lack of standardized codes for construction management approaches and paucity of vocational training courses related to upskilling of construction professionals has been observed.

The massive growth in the field of construction sector involves the development of more complex projects, requiring efficient advanced project management approaches which eventually require skilled project managers to lead project teams (Jailane Atef Amer, 2020). Hence, organizations tend to look for employees with efficient leadership skills and analytical abilities. (PMI, 2010)

The number of clearances needed may involve multiple agencies from central to state level and can take several months for obtaining approvals. To summarize, the major challenges found in the Indian construction industry are low mechanization, labour intensiveness, lack of technical skills, unstructured industry, and lack of contractor and technology experience.

1.5 Process of construction project management followed in India

There is a marked shortfall in competence and lack of capacity building in the construction industry in India. The increase in management professionals within the field of construction is insufficient to overcome the inertia and sluggish pace of progress within the industry. There is a dearth of standardized processes or references to PMBOK and other similar standards within projects in India. While IS 15883 (with 11 parts) has been established by the Bureau of Indian Standards, the guidelines are currently not sufficient for handling the growing project complexities and advancements in construction techniques which in turn are likely to impact the project. The appointment of a CPM, further, is devoid of standards for competencies, performance, and deliverables, which are accepted across organizations for evaluation of the performance of CPM. This is crucial especially for identifying the non-performance of CPM.

The concept of reactive performance versus proactive performance is widely used to evaluate approaches to project management. Project planning is critical during the project lifecycle, and a reactive project manager is characterized by a lack of comprehensive planning and rather responding to project risks as and when they arise. Consequently, such a managerial approach is likely to fail especially during the execution phase. In contrast, the proactive managerial approach ensures detailed planning prior to project execution that includes all knowledge areas based on the PMBOK (from the PMI) and also caters to the issues raised by the stakeholders. In comparison to reactive project managers, proactive project managers focus on the long-term interests and demands of the stakeholders associated with the project. Such managers have been observed to typically have good relationships with the project sponsors and other donors/beneficiaries of the project, which further ensures their efficiency. They are actively involved in all project activities and keep stakeholders updated throughout the process. Proactive project managers are known for being deeply involved in their projects. Proactive project managers keep a track of the risks involved during the crucial controlling and monitoring phase and update the risk register regularly.

This raises two important questions: What are the performance criteria of the CPM? How can the performance of the CPM be improved?

1.6 Validation of need for construction management performance

As per Project Management Institute (2021), project management refers to the application of tools, knowledge, skills, techniques, and resources on project activities for the delivery of intended outcomes.

A project may be well conceived and financed with the best talent pool of technical resources in terms of their expertise and knowledge, but due

6 *Construction project manager: role and practices*

to lack of proper coordination it might lead to poor project performance. Hence, a technically equipped and skilled resource in the management of the overall project is important to ensure the timely delivery of project objectives within the approved budgets (Udo Nathalie & Koppensteiner Sonja, 2004). The lack of understanding among project stakeholders about the multivariate nature of construction projects and its linkage with project success requires a project manager to define the complete project management plan for the entire project lifecycle (Unegbu et al., 2022). Construction projects tend to offer project managers with recurring challenges and uncertainty.

The four major values which are considered to be of utmost significance for the project management community as per the PMI Code of Ethics and Professional Conduct are:

- Responsibility,
- Respect,
- Fairness, and
- Honesty.

Construction projects are highly structured endeavours involving different sectors and requiring expertise of a variety of disciplines throughout their life cycle right from the inception stage. Projects in the construction industry turn out to be unique when compared with other industries and therefore require different management processes and practices (Park et al., 2017). In developing countries, the performance of construction industry and its projects has typically been observed to be retrogressive and hence requires tracing of the root cause, which is significantly related to lack of rigorous control over the project performance measures (Chan & Chan, 2004).

Essentially, the term construction management refers to the management of construction projects. With increasing complexity, innovative technologies, and emerging construction materials, projects tend to develop uncertainties and unpredictability in terms of their functioning and peculiarities.

Considering the dynamic nature of the construction industry, it would be implausible to conduct a construction project without a project manager. As the size and complexities of projects increase, the project management functions tend to become more critical. Hence, a CPM's role is crucial in ascertaining the success of a construction project (Udo Nathalie & Koppensteiner Sonja, 2004). Further, the growing need of human resource management and key performance measures in the construction industry can be seen as one of the most viable ways to define the excellence of a CPM (les Pickett, 1998).

The existing literature also demonstrates that in the case of extremely prudent clients who may not be satisfied with the fulfilled project objectives after complete handover of the project, it is the project manager's skills, competencies, and leadership that influence the success of the project and fulfilment of the clients requirements. Therefore, it is important to define the criteria for assessment/evaluation of a CPM's performance in order to achieve project goals and client satisfaction (Faisal Alqahtani et al., 2015).

Through literature review, it has been noted that there is a lack of evidence for a very specific systematic study related to understanding the taxonomy of competencies of CPMs. Project managers have been observed to lack the knowledge that can help propagate their continuous professional development towards best practices related to project management and strive for better performance. As identified by the researchers, competency-based approaches are a viable option for validating the performance assessment of an individual (Cheng et al., 2005).

Researchers and industry experts have also highlighted the importance of performance assessment as it engenders an individual towards enhancing their body of knowledge through skill development. Hence, it is becoming increasingly important to develop methodologies for the evaluation of the performance of a CPM (Bourne & Walker, 2006; Tunji-Olayeni et al., 2014).

As per Project Management Institute (2017), project management deals with the application of skills, knowledge, tools, and techniques on project activities in order to meet project requirements for the delivery of projects as per the client expectations. This requires the CPM to develop skill-set relevant to project management functions and processes. Lack of adequate project management skills can lead to cost overruns, missed deadlines, rework, poor quality, uncontrolled expansion of projects, loss of organization's reputation, stakeholder dissatisfaction, etc. In order to maintain continuity in activity and recognition in the current economy, companies are embracing project management services to ensure constant delivery of business value.

Efficient project management should be a part of the organizational strategy in order to achieve the organizational goals. It helps an organization in the following ways:

- Linking of project results with organizational goals
- Competing more effectively with peers in the industry
- Enduring the organization
- Developing adequate responses to market (Project Management Institute, 2017)

Measurement of construction project performance is an essential part of project management processes and control decisions and must be carried out in a systematic manner.

1.7 Prevailing practice of project management in construction

The prevailing practice of project management in construction is governed by the involved project stakeholders. The stakeholders in a construction

8 Construction project manager: role and practices

project include the consultants onboard such as but not limited to: architect, structural designer, sustainability consultant, landscape consultant, owner, and construction agency. As per BIS (2009), stakeholders are the individuals/organization who have interest in the success of the project. The stakeholders might also vary with the project delivery method adopted based on which the organizational structure is defined to facilitate the constitution of different agencies involved in the project. An example of organization matrix structure for design bid build model from IS15883 (Part 1)2009 is represented in Figure 1.

As illustrated in Figure 2, each consultant (Architectural consultant, structural consultant, MEP consultant, landscape consultant, etc.) involved in a construction project has his/her project manager designated for the project who is addressed as a project participant. These representative project managers are responsible for the success of project with a keen interest

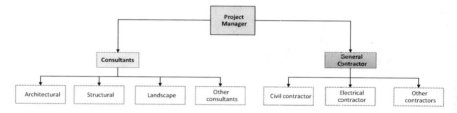

Figure 1 Typical organizational structure of design bid build model.

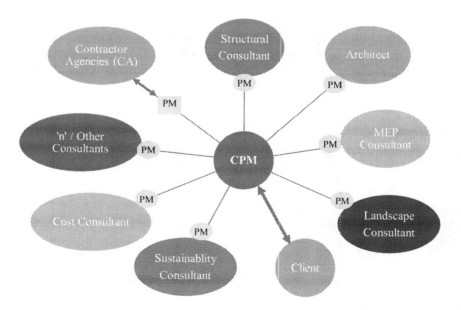

Figure 2 Practice of project managers in organization hierarchy.

Construction project manager: role and practices 9

towards their own organizations instead of the Client's interest. Similarly, a project manager is deployed by the construction agency as well for the execution of the project. These project managers of all the above-mentioned individual entities/stakeholders involved in the project are responsible for achieving the project objectives as per the agreed scope of work. These individual project managers of all entities report to the CPM who represents the client.

Figure 2 represents the project manager of the contractor agency. He/she is the strongest entity amongst the project members of different stakeholders affecting the performance of the project.

Please note that the number of stakeholders might vary based on the project requirements and the type of project delivery model selected for the project. Further, there might be scenarios requiring consultants and team of subcontractors.

CPM is the project manager reporting to the client/owner of the project.

> The CPM discussed in the concept of VDPI discussed within the book is indispensable part of the project and the full custodian of client's interest, overseeing the performance of the project on client's behalf.
>
> Generally, the communication processes of the project managers of the consultants are either absent or are informally established on case-to-case basis.

The ideal case of project manager as discussed in this book for developing the VDPI assessment tool is the one at the central topmost level of hierarchy in the management organization governing other subordinates at other organizational levels as illustrated in Figure 2. Based on the number of stakeholders (S1, S2,...Sn) and project requirements, different interaction mechanisms would be established during the project lifecycle, which will lead to the development of emerging complexities (Complexity 1, Complexity 2,...,Complexity n). Through the interactions of these interfaces and based on these complexities emerging through the stakeholder interaction, the level of project management skills needed to be utilized for the management of the project is established and represented in Figure 3. So, the utilization of project manager's level of competence would be specific to the project. These evolving complexities might also vary throughout the phases of the project lifecycle, so the level of competence needed would be a function of project typology, size, and phase of project at which the project manager is being appointed for the project. So, the objective of deployment of the project manager would be grounded by the project-specific characteristics or the emerging complexities.

The knowledge and experience of the client also govern the required level of competence of the CPM. The degree of interaction of the client throughout the project governs the depth of assessment undertaken to evaluate the competence of the project manager as described in Figure 4. Hence, the

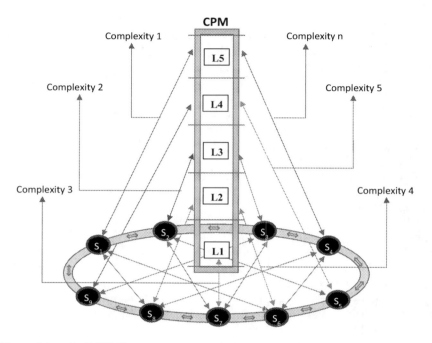

Figure 3 Level of CPM's competence needed based on the emerging complexities out of stakeholder's interaction.

level of competence of a CPM required can be described as a function of the degree of involvement of the client and is further governed by the client's knowledge and experience in the given field of the project for which the CPM is to be deployed.

The project manager's role in government organizations and private organizations varies as a private organization allows the project manager with greater flexibility of application of a new technology in comparison to a government organization. In the context of Indian construction industry, the two terms can be coined: *CPWDization* and *PSUization* in the government sector. The two terms can be applied to general practice of defining a project manager as the Engineer in-charge, who is too rigid to adapt to new technologies and international standards and overlook the need of enhancement of their existing codes related to the competence of CPMs. They are characterized by their use of the traditional approach of project management such as the Central Public Works Department (CPWD) works manual which, in practice, may not be the adequate criterion for defining the role and responsibilities of the project manager. In the manuals, it is clearly mentioned that these manuals are being prepared for the organizations (i.e., CPWD) use and can be used by other organizations at their own discretion

Construction project manager: role and practices 11

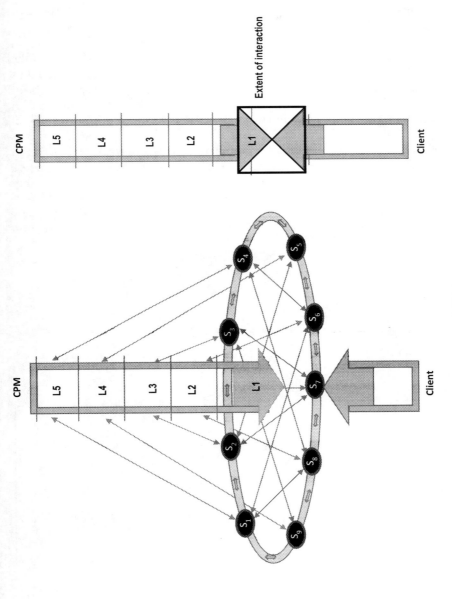

Figure 4 Extent of client interaction and level of competence of CPM.

but the major problem is that all other organizations, especially the Public Sector Units (PSUs), are blindly following these manuals which may not be suitable to the project conditions and there is a lack of defined criterion for assessing the performance of project manager.

Conceptually, the major intent of the VDPI theory is the establishment of objectivity of the CPM.

In general, the number of stakeholders governing a project is based on the project delivery model chosen. This also governs the formulation of the organizational matrix as represented in Figure 5. The client is depicted as the major stakeholder of the project who appoints a project management consultancy that acts as the client's representative to other agencies and consultants of the project. Further, the CPM is an employee of the project management consultancy and is responsible for governing other consultants and contractor agency. The latter is often termed as a general contractor and is further involved through engaging relevant subcontractors or nominated contractors by the client. There is a dedicated project handover team that operates during the final phase of the project. The figure is a generalized representation and variations in the organizational hierarchy model are likely to exist. For instance, in case of Engineering, Procurement, Construction

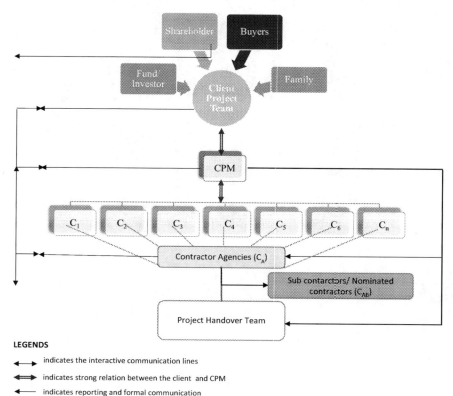

Figure 5 General organizational hierarchy matrix representing stakeholder interaction.

Construction project manager: role and practices 13

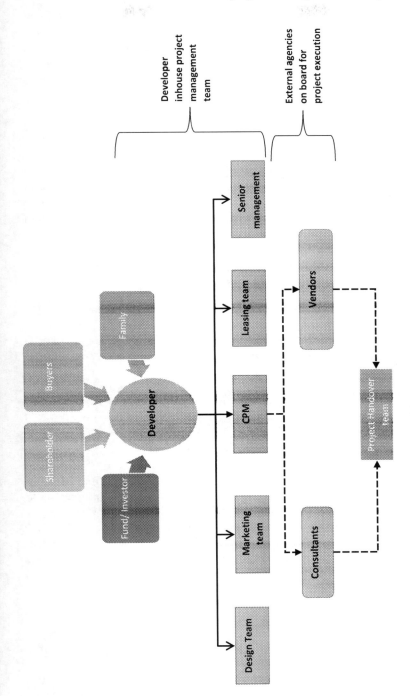

Figure 6 Developer-based organizational hierarchy model.

14   *Construction project manager: role and practices*

(EPC) projects with a single consulting agency bearing the responsibility of executing the project. There is also a general tendency within the construction industry involving the ad-hoc use of the project manager's nomenclature by all stakeholders involved in the industry. For example, a general contractor has a designated CPM. Similarly, an architectural firm also assigns a project manager responsible for carrying out the project deliverables.

The developer organizations have a distinct project delivery model. In such models, the client's in-house teams are appointed to manage the project throughout the project lifecycle till project closure. These teams may either work with vendors directly during the project execution or are deployed as consulting managers to report project updates to the client and their respective project teams. This structure of the developer organization-led project management has been encapsulated within Figure 6.

The authors argue that the implementation of traditional approaches of management within large-scale projects is now obsolete with the advent of newer methods of project delivery. Further, the evolution of these delivery methods requires a management approach that adapts to the changes within the industry and within global practices.

1.8 Inferences

Inferences have been derived through the existing literature cited and referred to within the above discussion. The challenges related to construction project management have been identified that focus on the industry practice for defining a CPM, nature (reactive and proactive) of a CPM and how it governs the success of a project, and how critical is the performance of a CPM for successful project management. This has been established based on the discussions carried out within the chapter.

The identification of the best talent pool for delivering a project, considering the uniqueness of every project, is necessary to deal with the crisis of skilled and technical manpower requirements. This is also crucial as construction projects tend to offer project managers continuous challenges and uncertainty.

The vague terminology used in the field of construction within India has also emerged as a major issue, since individual organizations have their own terminologies established for defining a CPM, but as a CPM an individual is responsible for the successful delivery of the project which makes it essential to define a standardization of performance expected from a CPM.

References

BIS. (2009). *IS 15883-1 (2009): Construction project management - Guidelines, Part 1: General.*

Bourne, L., & Walker, D. H. T. (2006). Visualizing stakeholder influence--two Australian examples. *Project Management Journal, 37*(1), 5–21. https://www.pmi. org/learning/library/visualizing-stakeholder-influence-stakeholder-circle-2558

Chan, A. P. C., & Chan, A. P. L. (2004). Key performance indicators for measuring construction success. *Benchmarking, 11*(2), 203–221. https://doi.org/10.1108/14635770410532624

Cheng, M. I., Dainty, A. R. J., & Moore, D. R. (2005). What makes a good project manager? *Human Resource Management Journal, 15*(1), 25–37. https://doi.org/10.1111/J.1748-8583.2005.TB00138.X

CMAA. (2022). *What Is Construction Management? | Construction Management Association of America.* Retrieved June 22, 2022, from https://www.cmaanet.org/about-us/what-construction-management

Construction | Make In India. (2022). Retrieved June 22, 2022, from https://www.makeinindia.com/sector/construction

Faisal Alqahtani, E., Chinyio, E., Mushatat, S., & Oloke, D. (2015). Factors effecting performance of projects: A conceptual framework. *International Journal of Scientific & Engineering Research, 6*(4). http://www.ijser.org

Jailane Atef Amer. (2020). Challenges in construction project management as faced by millennials in developing countrieas. *Jailane Atef Amer, 9*(7). www.pmworldlibrary.net

les Pickett, H. R. I. (1998). Competencies and managerial effectiveness: Putting competencies to work. *Public Personnel Management, 27*(1), 103–115. https://doi.org/10.1177/009102609802700110

Park, K., Lee, S., & Ahn, Y. (2017). Construction management risk system (CMRS) for construction management (CM) firms. *Future Internet, 9*(1), 5. https://doi.org/10.3390/fi9010005

PMI. (2010). *Project Management Practices in India-2010.* Report by Indicus Analytics and Ace Global.

Project Management Institute. (2017). *A guide to project management body of knowledge (PMBOK guide): Vol. Sixth edition.*

Project Management Institute. (2021). *The standard for project management and a guide to the project management body of knowledge (PMBOK guide).* (7th ed.).

Tunji-Olayeni, P., Mosaku, T. O., Fagbenle, O. I., Omuh, I. O., & Joshua, O. (2014). Evaluating construction project performance: A case of construction SMEs in Lagos, Nigeria. *Vision 2020: Sustainable Growth, Economic Development, and Global Competitiveness - Proceedings of the 23rd International Business Information Management Association Conference, IBIMA 2014, 1*, 3081–3092. https://doi.org/10.5171/2016.482398

Udo, N., & Koppensteiner, S. (2004). Core competencies of project manager. *Global Congress- EMEA, Prague, Czech Republic, Newton Square, Project Management Institute*, 1–7.

Unegbu, H. C. O., Yawas, D. S., & Dan-asabe, B. (2022). An investigation of the relationship between project performance measures and project management practices of construction projects for the construction industry in Nigeria. *Journal of King Saud University - Engineering Sciences, 34*(4), 240–249. https://doi.org/10.1016/J.JKSUES.2020.10.001

2 Performance assessment of construction project managers

Practices and challenges

2.1 Introduction

This chapter discusses about the contemporary industry standards used in assessing the performance assessment of a CPM, such as 360-degree feedback tool, project manager competency standards by Project Management Institute (PMI), Australian Institute of Project Management (AIPM), etc. The limitations related to these standards have also been discussed in the chapter regarding their criteria for assessing the CPM's performance. Other sections covered in the chapter are related to barriers in project manager performance assessment, solutions through the establishment of a performance assessment system, performance vs skills, and roles and responsibilities considered to be necessary for a CPM. The essence of the chapter lies in the value addition to the existing body of knowledge of project management practice through CPM performance assessment.

2.2 Limitations of existing standards

While several competency assessments and enhancement-related frameworks have been developed by researchers and construction industry professionals and are being widely used in practice, a lack of evidence in the literature has been noted regarding a specific systematic study intended towards understanding the taxonomy of the performance assessment of CPMs. However, project manager skills have been described in the existing literature as factors affecting project performance. To this extent, CPMs often lack the knowledge that can help propagate and sustain their continuous professional development towards achieving better practices in the construction industry. The traditional approaches of measuring performance typically encompass competencies related to time, cost, and quality. However, the constraints of time, cost, and quality are not particularly agreeable to isolate the inputs of project managers from other members of the team, in addition to the influence of extraneous effects (Ahadzie et al., 2008). The available standards are found to be performance-based and attribute-based for examining any project manager in general (GAPPS, 2007).

DOI: 10.1201/9781003322771-2

Performance assessment of construction project managers 17

Nijhuis et al. (2018) in their study have highlighted various challenges in the existing body of knowledge, such as a lack of homogenous list of competencies, necessitating a taxonomy, and the utility of importance as a criterion, which favours general important competencies.

The 360-degree feedback tool, also known as 360-degree assessment, has significant potential to identify the performance of an individual through their feedbacks, often taken from peers and employers. Despite the fact that 360-degree evaluation has become increasingly popular, its value has been questioned and challenged for applicability and relevance. Based on user experiences, certain drawbacks of 360-degree assessment have been identified:

1 Performance benchmarks and indicators are not considered to evaluate the performance level of employee.
2 Determinants of those indicators are not recognized in the assessment tool.
3 Authenticity of feedback can't be validated.
4 Respondents may focus on the weakness side and overlook the strengths.
5 Lacks of evidence in case of organizations that have low level of trust.
6 It depends on the capability to generate reliable data from unreliable sources.
7 Time-consuming task.

Therefore, since the organizations are keen to support the professional development of their managerial staff, an evaluation system that focuses on the criteria for project success outcomes can prove to be useful to assess and improve the performance of individuals and teams. The existing standards and models are being used to assess the performance of individuals and organizations; however, they lack in terms of background theory and adopted methodology. While organizations execute projects and monitor the performance of the project managers regularly, evidence of agreement in the literature on the methodology or process for evaluation of their performance specific to project performance is still missing. As identified by Zwikael and Meredith (2021), existing project success evaluation models cannot be applied to all project typologies, in addition to a lack of separation of project success measurement from that of project individuals' performance (for instance, of the project manager). The role of a project manager is multifaceted and significantly impacts the project success. A lack of agreed-upon set of roles and responsibilities of a project manager has been observed in literature, and the range of responsibilities for a project manager varies from administrator to multimillion budget manager (Udo & Koppensteiner, 2004).

The onus, therefore, is to suggest a set of evaluation criteria that incorporate the aspects related to project performance and individual domain of performance. The measures entailed need to be segregated into contextual

18 *Performance assessment of construction project managers*

and task performance behaviours of individual as well as projects and are linked with organizational goals. Even as there is a hefty expanse of literature and project management standards for assessing performance of a construction project, with multiple tools developed by organizations for generic performance assessment of a project manager, these tools are specific to performance assessment of a CPM.

This gives rise to a need for the development of efficient project performance measurement plans and systems which would help in enhancing the efficiency of the organization as well as at an individual level, which in turn indicates the need for developing better tools and techniques for performance management (Cooke-Davies & Arzymanow, 2003). To summarize, it can be said that there exists a need to assess the performance of a CPM considering the criticality associated with the construction projects based on which a standard quantitative criterion for evaluation of their performance is required to be established.

2.3 Practice of CPM in private sector in India: who, where from, and what he/she does

The concept of construction project management has become widely acceptable in the Indian context, with the utilization of services of project managers prevalent both in public as well as private sectors. However, the application of project management is different in private and public sectors. The difference in the two sectors identified is majorly in terms of infrastructure, funds, implementation, requirements, scope of work, organizational support, and culture. In case of private sector, the role of CPM is clearly defined based on the project requirements. In private sector, typically the CPM has a bachelor's degree as a minimum qualification. Further, the CPM will have studied management, with training on-site. It is unlikely for most construction companies to hand over the responsibilities of a project to an individual with little or no experience of managing a construction site. Some of the certifications that the CPM is required have been listed below:

- Project Management Certification (PMP)
- National Council of Examiners for Engineer and Surveying (NCEES)
- OSHA training for safety guidelines
- Green Business Certification (LEED)
- The American Concrete Institute (ACI)
- Crane Operation Certification
- Aerial Lift Training
- APM Project Management Qualification by RICS (Royal Institute of Charted Surveyors)
- Construction Management Association of America – Certified Construction Manager (CCM)

A construction project entails investment in terms of finances as well as time, effort, and competence. Hence, the numerous variables and components involved in the roles and responsibilities of a CPM need to be examined in-depth.

2.4 Practice of CPM in public sector in India: who, where, from, and what he/she does

The public sector is yet to adapt to the advanced project management approach. A major issue identified in the public sector is the appointment of CPM as an engineer-in-charge by most of the organizations, PSUs, CPWD, State PWDs, etc. CPWD being one of the premiere construction agencies of the Government of India provides comprehensive building construction management services, with project being the responsibility of the department.

However, due to delays, management practices and bureaucracy observed in such agencies, decisions and data needed for an effective project management system seldom arrive to the project management team on time. Application of project management systems and processes in the private sector has been found to be significantly simpler.

As per the CPWD works manual, the engineer-in-charge is defined as "the officer entering into an agreement on behalf of the President of India or his/her representative responsible for execution of the work". The skill sets administered by the employees found to not meet the requirements of complex projects adequately, since the individuals often lack in terms of technical expertise. Further, they are made accountable for the project's performance. However, with the absence of skill sets and training, it is difficult to transfer responsibility to the professional appointed. Alternatively, a Catch 22 situation arises when there are no yardsticks for performance, and it is complicated to define the competencies.

Risk management and resource management are often overlooked areas in project management. Consequently, superior project management tools and practices involving lean thinking, Building Information Modeling (BIM) applications, technological advances of 3D printing, deep learning, data analytics, AI, robotics, etc. may not find their way into project management systems in the Indian context, unless the performance is treated objectively rather than vaguely ad-hoc.

Hence, VDPI as introduced in this book is an attempt to break this conundrum by introducing a performance framework that uses performance and process clarity to ensure that the competencies can be cultivated.

2.5 Existing body of knowledge for competence development

Competency can be defined as the "possession of sufficient knowledge or the ability to do something". It can therefore be said that the skillset of a project manager is critical for accomplishing a successful project.

20 *Performance assessment of construction project managers*

Lynn Crawford's (2000)[1] definition of competence is, "Competence is a term with different meanings for different people. But it is generally accepted as something that encompasses knowledge, skills, attitudes, and behaviours that are causally related to superior job performance".

Major components of competency include (Cartwright & Yinger, 2007):

- Abilities
- Attitudes
- Behaviour
- Knowledge
- Personality
- Skills

Project management competence is defined as the ability to integrate technical skills, cognitive aptitude, and interpersonal skills in order to achieve project objectives (Bashir et al., 2021). A competent project manager incessantly poises five projects " currencies" – time, money, knowledge, security, and prestige and is the best possible mix of the knowledge, performance, and personal competence (Udo & Koppensteiner, 2004).

Various standards relevant to competency assessment have been formulated by major international organizations related to project management such as the PMI and AIPM. Some of the tools developed by researchers and academicians include Leadership Dimensions Questionnaire (LDQ), the Mayer-Salovey-Caruso Emotional Intelligence Test, the McBer Competency Framework, the Inwald Personality Inventory (IPI), the General Mental Ability (GMA), the Myers-Briggs Type Indicator (MBTI), and the Multifactor Leadership Questionnaire (MLQ) to assess project managers' competencies (Bolzan De Rezende & Blackwell, 2019).

2.5.1 *Project manager competency development framework by PMI*

The project manager competency development framework provides the definition, development, and assessment for any project/program/portfolio manager competencies. It defines the key competency areas to be considered and their impact on overall managerial success. Additionally, the framework evaluates the performance of a project manager using three key competency areas: knowledge, performance, and personal competencies as represented in Figure 7.

Knowledge competence – it is assessed by completing relevant examinations and gaining their credentials/accreditation like Certified Associate in Project Management (CAPM), Association of Project Management

---

1 Crawford, L. (2000) Project management competence for the new millennium. In: Proceedings of 15th World Congress on Project Management, London, England, IPMA.

(APM), Program Management Professional (PgMP), Project Management Professional (PMP), and Portfolio Management Professional (PfMP).

a   Performance competence – it is measured using the project performance of the desired deliverables and produced outcomes, which is based on their knowledge and skillset.
b   Personal competence – it is evaluated on the basis of individual's behaviour. Units of personal competence are defined as leading, communicating, professionalism, managing, effectiveness, and cognitive ability.

The standard also briefly provides insight into the path of development for a project manager from the novice level to the experienced level based on the gained experience and improvement of competencies in terms of both technical and soft skills.

The competency development process as described in Project Manager Competency Development standard by PMI can be applied either incrementally or holistically to set of competencies subject to individual, project, or organizational requirements. The following steps as represented in Figure 9 are involved in the competency development process (Project Management Institute, 2017c):

• Review of requirements – to identify the gap, goals, and needs based on which the criteria of assessment is established
• Assess competencies – to identify the areas of strengths and weaknesses
• Prepare competency development plan
• Implementation of developed competency development plan

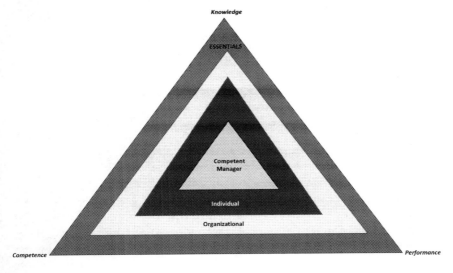

Figure 7 Project manager competency development model.

22 *Performance assessment of construction project managers*

It also tries to establish an assessment criterion for each target group (Hanna et al., 2016). In a study by Udo and Koppensteiner (2004), knowledge competence is defined by three pillars of knowledge as described below:

i   First pillar of knowledge – The first pillar consists of skills such as negotiation, leadership, team building, communication, and other human resource management skills related to general management.
ii Second pillar of knowledge – The second pillar consists of knowledge related to tools and techniques used in the areas of project management like cost, scope, time, quality management, etc.
iii Third pillar of knowledge – The third pillar comprises industry-related knowledge like life cycle management, product development methodologies, etc.

Further, personal competence can be divided into six areas – helping and human service, achievement and action, impact and influence, cognitive, managerial, and personal effectiveness (Bashir et al., 2021). Personal competence is generally described by considering two sets of competence – personality characteristics and resource management skills. The indicators are related to personal competence and job focus, such as team building, leadership, maturity, honesty and integrity, approachability, decision making, learning understanding and application, communication, adaptability, attitude, enthusiasm, and self-efficacy (Oliveros & Vaz-Serra, 2018).

Performance competence is described by the project manager's experience in the industry, hours of exposure to project management practice, track records, complexity and size of projects managed, etc. (Project Management Institute, 2017b).

The PMBoK guide to project management identifies five areas for defining the competencies of a project manager. These competencies have been grouped with the knowledge areas of the PMBoK guide as follows:

- Project management application
- Understanding of the project environment
- General management
- Technical area expertise
- Interpersonal skills

The role of a CPM requires various skillsets such as interpersonal skills related to communication, leadership, problem solving, negotiation, team building, decision making, management, etc. and technical skills related to the expertise for project, from inception to handover stage of project life cycle. In recent studies, construction and engineering organizations have been found to exhibit better performance capabilities and high levels of maturity towards the achievement of project goals, which is a result of information sharing, leadership, and degree of authorization (Zwikael & Globerson, 2006).

Even with the dynamic nature of evolving projects, technologies, and a profound change in the industrial paradigm, there would always be the need of a unique combination of human cognizance and the key elements of project management. To comply with the changing trends in the construction industry, competencies of project managers need to be updated and evaluated for better performance of the managers as well as projects (Ustundag & Cevikcan, 2018).

Performance measures are used to define the managerial excellence of a project manager. Some of these are achievement orientation, information seeking, initiative, directiveness, focus on client's needs, cooperation, and teamwork. An example of indicators to be considered in case of a construction project is represented in Figure 8.

The indicators which are considered for the time, cost predictability, safety, client satisfaction, productivity, and profitability are related to the project's performance and are result-oriented, except for the predictability of cost and time for design and construction, which are more oriented towards procurement and safety. These performance indicators are used to identify the performance of projects. However, there are very few indicators to measure the performance of stakeholders, for which there is a growing need throughout the project phases (Takim & Akintoye, 2002).

Based on our research study, five major indicators with respect to a CPM's performance have been identified. These are – scope, time, design, contract, and cost. These have been linked to the three pillars of a project manager's competency, which are knowledge, experience, and personality. These three pillars describe the main areas of a project manager's individual competencies related to project performance.

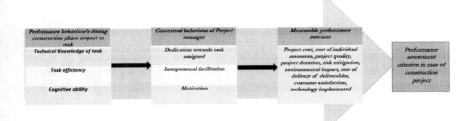

Figure 8 Project manager performance assessment criterion example in case of construction project.

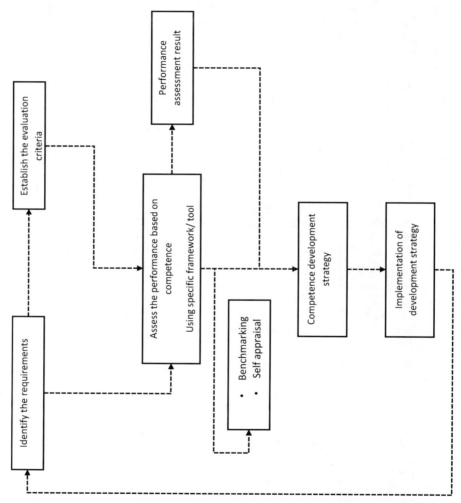

Figure 9 Competence development process.

2.5.2 *Association of project management competency framework*

The APM Framework lists down the activities crucial for effective project management. It tries to mention both good practices as well as needs that will arise during the execution of the project. APM Framework also establishes a common benchmark to be achieved for individuals and organizations with respect to project activities. It comprises of 27 competencies that are associated with the deliverables using a 5-point scoring scale for the assessor to evaluate one's competency. The self-assessment form is based on evidence for an individual scoring.

2.5.3 *Australian Institute of Project Management (AIPM) professional competency management standards*

The standard developed by AIPM provides the basis for the assessment and development of a project practitioner. Being a generalized standard, it can be applied on project practitioners belonging to different industries/ enterprises. The major units that govern the professional competency management standard are scope, time, cost, quality, human resource, communication, risk, contract, and procurement management techniques. Each unit comprises of its own set of indicators, assessment performance criteria, knowledge skills, and evidence guide.

Range indicators for each unit have been mentioned in Table 1.

In the context of the construction industry, there is a growing need of human resource management, and key performance measures act as one of the most viable ways to define the excellence of a manager. Ahadzie et al. (2008) introduced the multidimensional competency-based conceptual model and also argued that the traditional performance measures are not sufficient to gauge a project manager's performance, as they lack in providing the appropriate information required towards stimulating their continuous professional development.

The unique nature of construction projects also governs these performance measures and their level of relevance in gauging the project manager's performance. The actual performance gap for the project manager is the corollary of the evaluation carried out using these indicators.

2.5.4 *Complex project manager competency standards*

The standard for complex project manager competency lays down the guidelines essential for the delivery of complex projects. It is divided into two sections; section 1 covers the underlying standards based upon the work of (Dombkins, 2007) called Complex Project Management. Section 2 specifies the knowledge, competencies, and special attributes for traditional and complex project manager. It is based on the concept of system thinking which considers project management approach to be holistically dealing

Table 1 Range indicators for each management unit

Management unit	Range indicators
Scope management unit	
Contribute to scope definition	Contributing to the identification of project deliverables
	Supporting the establishment of the process of lifecycle management for the project
	Contributing to the identification of criteria for project acceptance
Apply project scope controls	Undertaking the work according to the agreed project management plan and/or agreed business plan in order to support efficient change control and procedures for performance measurement
	Monitoring the assigned compliance areas related to the scope requirements and communicating shortfalls to the project manager
	Measuring progress in order to determine the perceived, potential and actual scope issues which may need formal change in scope
	Contributing to recording and reporting of scope changes as agreed with the stakeholders within assigned work responsibilities
	Supporting the application of project reporting and monitoring systems so as to enable the performance evaluation of the project
	Assisting in the assessment of project outcomes in order to determine the efficiency of initial and subsequent approaches employed for scope management
Time management unit	
Contribute to the development of project schedules	Supporting the identification of sequence of activities, dependencies of tasks, duration of completion and effort required, in order to meet the project objectives as assigned
	Contributing to the development and establishment of the Work Breakdown Structure in the context of project schedule development, which also includes consideration of risk and estimation of impact(s)
	Identifying the impact of schedule on cost estimation and risk characterization
	Contributing to the development of the project schedule management plan
	Supporting the introduction of tools and techniques for planning and scheduling of the project, which is required for other time management aspects of the schedule

Monitor agreed schedule	Recording and reporting of variance between planned and actual progress on allocated tasks within the project schedule
	Contributing to the identification of tasks that may be integral to the critical path(s)
	Supporting processes for monitoring in order to identify the deviations from the schedule that may have an impact on achieving the project objectives
Update agreed schedule	Updating the plans and schedule as directed in order to accommodate changes throughout the project life cycle
	Using scheduling tools for measurement, recording and reporting of the progress of activities in accordance with the agreed project plan and schedule
	Contributing to the forecasting of impact of changes made in schedule and analyzing the options
Contribute to implementation of project schedules	Contributing to the assessment of progress against the schedule throughout the life cycle of the project
	Documenting project progress and changes in schedule as per the project documentation standards
	Monitoring the consistency of changes in schedule and aligning it with changes in objectives, scope, risks and constraints
Participate in assessing time management outcomes	Assisting in the assessment of project outcomes in order to determine the efficiency of scheduling tools, approaches and techniques used
	Identifying the time management and scheduling issues for application in future projects

28 *Performance assessment of construction project managers*

with the projects. The focus of the standard is to view a problem through multiple metaphors to obtain a better and practical understanding using system thinking. The complexities discussed in the standard apart from those derived from the PMBoK include engagement, leadership, assertiveness, self-control, openness, relaxation, result-oriented, creativity, consultation, efficiency, conflict, negotiation, ethics, values, and reliability. Some of the special attributes included in the standard are action and outcome oriented like wisdom, focus, courage, creativity that lead innovative teams and have the ability to influence others. In total, nine views are defined with respect to the role of a complex project manager: strategy and project management, system thinking and integration, reporting and performance measurement, business planning lifecycle management, innovation creativity and working smarter, change and journey, culture and being human, leadership and communication, and probity and governance. The standard has set three levels for classifying project managers based on their competencies – traditional project management, executive project management, and complex project management. Here, the actions needed in the workplace by the project manager are detailed out and grouped into traditional, exceptional, and complex project management.

2.5.5 *360-degree perspective from stakeholders*

The involvement of various stakeholders in a construction project, such as contractors, owners, and consultants, leads to a complicated relationship among different contracting parties. Wang, (2016) conducted a study on the relationship between project success and the stakeholders' performance, where it was proved that the supervisor, owner, and the contractor's performances are substantially associated with the criteria of project success. Hence, it becomes all the more important to measure the performance of various project stakeholders in order to assess the competence of an individual.

There are a plethora of assessment models to assess the performance of a project based on certain indicators, whereas assessing the competency of an individual is still a challenge for the Architecture, Engineering and Construction (AEC) industry. Assessments of an individual or self-assessment are indicators to evaluate or an opportunity to improve certain areas where performance has not been met with the set benchmark. Moreover, there is a growing demand for suitable assessment models for assessing an individual's performance in the AEC industry, so as to involve competent stakeholders in order to accelerate the economy of the construction industry ensuring the best quality standards.

360-degree feedback tool, also known as 360-degree assessment, can be defined as "the systematic compilation and critique of performance data for an individual or a group, which can be derived from various stakeholders in their performance". The individual can assess the challenges

Performance assessment of construction project managers 29

and competencies associated with the program, as it also facilitates better working relations and prepares for the next level of leadership. Delivering the strategy demands greater flexibility; it helps project managers in implementation of their own development road map as well as focusing on personal development. The formulation of the feedback tool is used in assessing the core competencies of the individual linked with the organizational goals (Kumar Das & Panda, 2015).

It has been observed that the best use of 360-degree feedback in ideal circumstances is seen for personal growth rather than evaluation (Tornow et al., 1998). As compared to single-rated techniques for feedback, 360-degree feedback has been proved to offer certain advantages. Instead of relying on feedback or perceptions of a single individual, multi-rated feedback helps collect multiple perspectives from various angles, which helps to provide a more complete and inclusive idea of the performance of the employee under scrutiny (Hosain, 2016). Further, as suggested by London and Smither (1995), 360-degree feedback can prove to be an effective organizational intervention and can help to raise awareness of the necessity of harmonizing behaviour, customer expectations, and work unit performance, as well as promoting engagement in work effectiveness and leadership training. It also recognizes the sophistication of management and the importance of data from various sources. The procedure helped the managers to comprehend the idea and fundamentals of competency models, and further associate it with their own performances. Competency models are also developed to assist the organization in achieving its objectives. Usually used for learning and development, 360-degree feedback has been observed to be more successful when combined with appraisal (Parker-Gore, 1996).

Jones and Bearley (1996) identified the primary reason for decision makers to seek feedback on their areas of expertise, which is that it offers knowledge on a leader's present actions and others' expectations; it acts as an assistance process for continual learning; it helps executives in the validation of the perceptions of themselves; it ensures that representatives have a realistic view of themselves; and, most importantly, it incentivizes management leaders to invest in the efficiency of their leading figures.

In terms of new work arrangements, the popularity of 360-degree feedback is also driving. Peer feedback has become increasingly important as hierarchies have compressed and more work is done across departments and cross-functional groups (Toegel & Conger, 2003).

The 360-degree feedback survey should emphasize relationship building in order to foster common understanding and mutual insight, as well as personal self to learn. Understanding group behaviour is complicated considering the fact that the leader is generally the one who directs the group toward its objectives. Consequently, an improvement in group performance can be expected following more precision in forecasting capability. The 360-degree feedback encourages individuals to examine themselves and

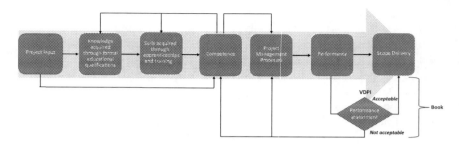

Figure 10 Performance and competence-based diagnosis.

focus on improving themselves, thereby achieving the organization's goal (Kumar Das & Panda, 2015).

2.6 Performance versus competence

While one generally appreciates the difference between words such as performance, competence, qualification, skill, and processes, there is a tendency to use these terms in synonymity. For instance, if there is a default in the discharge of responsibility, one can use any of these and "broadly" conjecture the diagnosis. Indeed, the default can be on account of any of these issues but when it comes to specific improvement action, one may catch the bull wrong way. Hence a clarity is essential for prognosis and outcome-oriented diagnosis. There is, however, a symbiotic relationship that cannot be ignored. At the same time, there is an order in which these terminologies manifest in project management practice. This can be understood in Figure 10.

Performance being the prime function of VDPI, the contrast and comparison between performance and competence are presented in Table 2.

2.7 Qualification

Qualification is the foremost requirement for the project manager. For any domain, it is necessary to have the domain knowledge acquired through qualification, typically a bachelor's degree at the very least. In the context of a CPM, appropriate qualification in architecture, civil engineering, or another engineering discipline that is closest to the project nature, must be acquired. Otherwise, a simple degree in MBA or BBA, as it prevails in many countries, may not provide a substantive edge on a construction project, especially when complexities begin to manifest. Further qualification in project management provides the necessary legitimacy for engaging in managerial decisions and positions. It is understood that in most

Performance assessment of construction project managers 31

Table 2 Performance versus competence

Performance	Competence
Activity	Ability to perform
Doing	Knowing
Evaluating performance	Assessed through evaluating performance
Measured against standard, defined or benchmarked	Encompasses skill and knowledge
Fulfilment of obligations	Underlined knowledge for a task
Compliance to contract	Idealized anticipation for a task
Affected by distractions.	Idealized intellectual property
Realization of action	Decision-making
Act of the system	Focus on the system
Manifestation of language	The language or syntax making a system.
Usage of the system	Constitutes a number of elements, making the system itself
Only focus during action	Predominant focus of preparation
Depends on perseverance and confidence to put into action	
Depends of memory, spot decisions, initiative	
Theory of practice	Theory of body of knowledge
Real world output	Subconscious understanding
Reflects competence	Acquaintance with rules, procedures
Represents a small set of knowledge and skill application	Understanding of dynamics of project and its morphology
Recognizing ambiguity and making sense out of it	Ability to recognize ambiguity
Not fault free mostly	Perfect
Performance determinants establish competence adequacy	Improvement in competence expects performance
Validation of competence improvement	Competence is a hypothesis
Decides robustness of assumption	Assumption
Let down	Cause of let down
It should decide threshold of competence required	It should decide threshold of knowledge required
	Suitable, sufficient
	Attribute based
Decides performance criteria	Defined on the basis of the inferred demonstrated ability for satisfaction of performance criteria
Typical skill and knowledge driven outcome-based employing processes	Project complexity driven outcome-based objective onboarding
Depends of complexity of situation	Assumes complexity

cases such management qualifications are acquired through experience which have a limited shelf life as the obsolescence of knowledge can render one unqualified soon. A formal education route to acquire knowledge ensures a comprehensive understanding on foundational principles that may not get outdated. A common practice to acquire qualification is by

32   Performance assessment of construction project managers

passing a professional examination, such as PMP, and, then keep updating through PDUs annually. Alternately, university education is an established approach. One can also argue in favour of experience-based qualifications as being adequate, in case of a particular organizational specialization. This may not be incorrect if the organization has a system of monitoring and developing human resources in terms of technical inputs as well as managerial qualifications from time to time. However, for higher-order challenges involving innovation, creative thinking, finance management, contracts and law, analytical strategic planning, and human behaviour, it is essential to follow a formal education route.

2.8 Skill

Complementary to knowledge acquisition in the component of skill. The proportion of skill necessary for a given project situation depends on the nature of engagement, organizational challenges, and experience of an individual. Skill without significant knowledge domain is adequate for a supervisory or superintendence role on the project. Skill can be divided into two broad categories, namely, task-oriented technical skills and soft human behavioural skills. Standard Operating Procedures (SOPs) are an integral part of formal task-oriented technical skill set.

The process of acquiring skills is different for task-oriented technical skills and soft human behavioural skills. Soft human behavioural skills require an aptitude and reflect innate personality of an individual. However, for a specific situation, such as those in a project, such skills can be inculcated to a certain extent which can be developed by an individual based on the motivation levels. It is however important to underline the fact that most of the managerial functions of non-contractual nature are performed through informal human interactions. Such skills are essential for the team building and a construction project manager must possess to discharge functions as a leader of the team. *Quality Management Systems, such as the ones propounded under the ISO 9001 family of system standards and globally recognized National Baldridge Quality Award system specifically emphasize on Leadership qualities that emanate out of soft skills.* However, for the purpose of VDPI narrative, the complication of soft skills is kept out of the scope.

Skill can be described on the basis performance framework and the acceptance criteria. Apprenticeship, training, simulation, or case study-based teaching are generally effective modes for skill development. Over a period, one acquires skill specific to the exposure but as soon as another technological context changes and performance standards are exacting, skill orientation or training becomes essential.

Skill requirement, as in case of qualification pre-requisites, depends on the complexity of the project. Complexity is a composite notion of expected challenges in each project that could include technological in-experience, risks in a project, and organizational experience.

2.9 Processes

In project management, standardization of process framework or protocols is very essential as a project manager does not function in isolation. Such process standards are established by professional associations or organizations such as PMI through Project Management Body of Knowledge. To create an eco-system of inter-organizational communication and seamless coordination amongst the stakeholders, such process frameworks are also adopted by IT-enabled tools such as Primavera. The purpose of such processes is to facilitate the application of knowledge and skill in a sync with the project development stages across all the stakeholder organizations. It is incumbent on the construction project managers to maintain the integrity of processes during discharge of their responsibilities. In absence of defined framework of processes, the application of knowledge and skill can only be ad-hoc and no assurance of success in a given project.

2.10 Competence

Competence is defined as aggregation of knowledge and skill. Competence determines how effectively the project management processes can be followed. On the other hand, processes complexity (based on the need of the project) determines the competencies necessitate. Competence and processes, however, cannot assure success unless put into practice. The action of putting into practice is understood as "performance", the core concept of VDPI. For further clarification, a comparison between performance and competence is discussed in the next section.

2.11 Barriers in performance

The traditional performance measures are not sufficient enough to gauge a project manager's performance, as they lack in providing the appropriate information that they need towards stimulating their continuous professional development. Further, the unique nature of the construction projects also governs the performance measures and their level of relevancy in gauging CPM's performance (Ahadzie et al., 2008).

A significant amount of effort is expended in the construction industry to measure the performance of projects with respect to cost and time indices. Further, the overall performance evaluation criterion as applied in the construction industry is found to be less structured in a subjective manner. Effective monitoring of construction projects comprises of two essential elements – quantification and integration of the different aspects of performance. Thus, it is of utmost importance for all organizations to measure performance since the improvement of process necessitates the measurement of success.

34 *Performance assessment of construction project managers*

Evaluation methodologies and procedures entail an elaborated mix of skills, knowledge, technologies, values, and routines. The lack of well-defined and consistent evaluation procedures and methodologies has led to arising disparities related to project judgments (Nassar, 2009). Thus, demanding better ways for performance assessment. The performance outcomes of an organization are determined by its effectiveness in the development and deployment of capabilities, including those necessary for project execution. As a matter of fact, it is a capability in itself to be able to develop and leverage new capabilities, which is called a "dynamic capability" (Bannerman, 2013).

The current literature identifies two concepts for defining barriers to performance as adapted from the literature "liabilities of incumbency" and "liabilities of newness" (Bannerman, 2013). Some of the challenges to effective project monitoring and control processes are listed as follows (Bohn & Teizer, 2009; Callistus & Clinton, 2016):–

- Weak organizational capacity
- Restrain in resources and limited budgetary allocations for evaluation and monitoring
- Deficiency of adequate resources and institutions that can address the constraints
- Ineffective linkages between budgeting, planning, evaluation, and monitoring
- Lack in demand for evaluation and monitoring, and its utilization, which further leads to inadequate quality, inconsistencies, and gaps in the data collected
- Diverse views, expectations of project stakeholders
- Most of the measures that are used for performance assessment can only report the performance after they have occurred (Callistus & Clinton, 2016)
- Limitation of processing information
- Lack of a detailed nationwide database for evaluation and monitoring systems, etc.

Measurement of project success is a challenge for any organization and turns out to be a very complex task. Generally, the organizations solely relate their organizational goals to the project's success, which is not the right criterion for setting up the performance evaluation system. The objective should be towards developing a robust performance evaluation criterion for construction practitioners which encapsulates the overall performance of an individual in terms of projects performance (Nassar, 2009). Another major barrier faced in performance measurement is a lack of knowledge about the key influencing factors, governing the project's as well as individual's performance. Sometimes, the determinants being considered for the performance evaluation in existing standards are found to be the same at all levels of the organizational hierarchy which might not always be true. Given that the perspectives of success vary for each project stakeholder and might not always be the same, evaluation models need to be more inclusive to be able to represent such differences in perspectives as well.

The traditional approach of project management is governed by the iron triangle of project management defined by project time, cost, and scope, based on the dynamic nature of construction projects the iron triangle is no longer valid to measure project management success and needs to be refined based on the organizational goals and project requirements (Atkinson, 1999). The performance assessment in the identified literature generally is defined using two scenarios: first being the final project outcomes and second being related to project management practices followed. A study by Meshram et al. (2020) indicates that the measurement of project managers performance is a very crucial task in case of construction projects and is based on a number of performance indicators involved and the data that needs to be collected and monitored on a day-to-day basis for tracking of the performance parameters to reach at the reasonable and acceptable levels of accuracy requires a standard approach.

The barriers faced in performance assessment can be triggered at various levels like organizational, project, and individual levels. Many organizations emphasize on the numerical value of the financial performance terms like return on investment and profitability index as criteria for measuring success which is not sufficient and a wholesome performance assessment framework is needed to address the gaps in performance (Rehman et al., 2012).

2.12 Way ahead through performance assessment

CPMs need to act on different levels in a project (Udo & Koppensteiner, 2004). It is the key onus of the senior management to identify appropriate competencies and success criteria that would help an organization in achieving a new paradigm of organizational performance and meet future demands of the growing markets, as its always about the survival of the fittest considering the present industry environment (Ahadzie et al., 2008). With the growing advancements and competition in the industry, how the projects are performing is the attribute defining organization's performance (Rehman et al., 2012). The use of performance assessment lies in the hands of the individual managers as well as the organization, as the identification of CPMs level of performance would help the better allocation of tasks to the project individuals and formulation of organizational strategies (Tunji-Olayeni et al., 2014). However, performance assessment does not guarantee performance enhancement straight away but acts as an indicator to identify the areas for improvement. The construction industry has a significant global impact on the global economy, hence performance measurement in the construction sector is extremely critical (Demirkesen & Ozorhon, 2017).

36 *Performance assessment of construction project managers*

Performance assessment is beneficial in terms of analyzing the focus areas for an individual or an organization. The chaotic and hard-to-adapt nature of emerging technology, in addition to the dynamic nature of construction projects requires enhanced management approaches. A project manager's competency assessment framework would identify the key competencies that are most likely to impact the performance of project managers in the respective fields of the entire management process (Project Management Institute, 2017b).

The aim of performance assessment is to evaluate the performance of an individual/organization which helps in developing the road map for growth; considering an unbiased evaluation procedure acting as a trigger for change based on the identified gap in performance and areas for improvement (Basu, 2015). Performance assessment is an integral part of performance management. It provides one with a framework or way ahead for performance improvement programme (Meshram et al., 2020). The existing studies try to measure performance using key performance indicators and link it with project success. However, there's no standardized way of performance assessment specific to the construction industry yet established (Demirkesen & Ozorhon, 2017).

Additionally, it nurtures the culture for performance enhancement. The approach of continuous performance assessment leads to clear identification of the primary objectives of an organization and how their accomplishment can be measured by the organization. It defines their efficiency and effectiveness in achieving the defined objectives. The concept of continuous improvement must be included in performance assessment as it eventually leads to the success of a wider range of projects impacting the organization's performance (Lannon, 2019).

Every project being unique in itself requires expertise with due diligence related to the aspect of project management. Nowadays, the competency level of a CPM is also gauged with respect to the overall project performance. As suggested by Oliveros and Vaz-Serra, (2018), there exists a positive correlation between project success and project manager's competencies. Major criticality lies with the management function of a project which is handled by the CPM based on his/her competencies. Few organizations see a project manager as an overhead resource but it is very essential to have a project manager in the construction industry considering the size, nature, and complexity associated with the construction projects. Thus, selection of a competent CPM is one of the most important decisions for any project's success.

2.13 Role, responsibility, and accountability

Construction management unlike general management profession is much more multidisciplinary. The construction industry practitioners have realized the importance of the project management concept for the successful handover of a project. A CPM has to develop professional management skill

Performance assessment of construction project managers 37

sets and use once abilities in the field in conjunction with their technical knowledge and skills, which can help in the standardization of processes to accomplish better results by reducing the associated risks using their own set of competencies. Establishing a criterion for assessing a CPM's performance is essential for defining his/her roles and responsibilities (Unegbu et al., 2022).

Managing a construction project typically includes identification of the project requirements in detail, keeping in mind the needs and expectations of the various stakeholders in planning and execution, establishing an effective and collaborative communication system among the stakeholders involved while focusing on the project deliverables, a project might be a temporary endeavour but the outcomes of the project may not be temporary and require a deep understanding of project constraints which include time, cost, scope, quality, risks, etc. (Project Management Institute, 2017a).

The following are the key knowledge areas in which CPMs are generally required to have a requisite set of skills in managing the areas of project management as per (IS 15883, 2015; Project Management Institute, 2017a):

- Project initiation
- Time management
- Scope management
- Cost management
- Risk management
- Communication management
- Human resource management
- Quality management
- Integration management
- Stakeholder management

It is the duty and obligation of a CPM to complete the project as per the client requirements; as stated in the project scope within the specific time and approved budget. A CPM leads the entire team for the successful delivery of the project and is responsible for shaping the process of project delivery. CPM's role involves:

- Elements of construction project management
- Clearly defined goals and objectives
- Ensures a comprehensive project management strategy for implementation
- Well-defined project management processes as per the team and project requirements
- A proven set of management tools
- Unbiased decision-making power
- Thorough understanding of the role of project management (Mohammed et al., 2016)

38 *Performance assessment of construction project managers*

Based on the project conditions, the CPM should also be equipped to perform beyond the scope of traditional project management (Udo & Koppensteiner, 2004) and must possess all necessary skills and competencies as needed right from the inception to the stage of occupancy (CIOB, 2014).

A CPM must be able to communicate with stakeholders involved at all levels of the business hierarchy; right from site labourer to the client and are required to deal with the challenges across the different phases of any project's lifecycle which have an impact on the project deliverables. Some of the techniques and project management processes are needed to be modified and made specific to the construction industry which requires a CPM, a manager is held accountable for daily project work right from inception stage up to handover stage of the project.

The duties of a CPM include but are not limited to the following hard and soft skills:

- Continuous duty of exercising and monitoring control over the project.
- Ensure professional, competent management co-ordination.
- Managing stakeholders
- Organizing, reporting
- Proactively disseminating project information to all stakeholders
- Innovation
- Decision making
- Well-defined construction plan
- Prioritization
- Team building
- Determines how construction work must be split into packages
- Management of overall site facilities: access, storage, facilities
- Supervision of execution work packages
- Constructability review

A detailed list of role and responsibilities of any project manager as per CIOB (2014) are listed in Table 3.

The responsibilities for any CPM vary from project to project (Ali & Chileshe, 2009) based on the scope of work, and type of organization; the list of responsibilities in Table 3 is a robust set of duties which a CPM is generally held responsible for as it varies with the size of project which directly has an implication on the project team members, like in case of a small fit-out interior project a single project manager is held accountable for all responsibilities of the project, whereas in case of large-scale projects the organization might appoint project managers for each major processes of project management like separate commercial manager, procurement manager, construction manager, monitoring and control manager, design manager, etc. based on the project requirements.

Performance assessment of construction project managers 39

Table 3 Responsibilities of a project manager as per CIOB

S.NO.	Responsibilities of a project manager
i	Be party to contract
ii	Assistance in preparation of the project brief
iii	Development of the project manager's brief
iv	Providing advice on site acquisition, site planning, grants
v	Providing advice on arrangements for funding/ budgets
vi	Development of project strategy
vii	Arranging the feasibility study and report
viii	Development of the consultants' brief
ix	Preparation of the project handbook
x	Selection of the team members in the project
xi	Devising the project programme
xii	Coordinating the design process
xiii	Establishing the management structure
xiv	Arrangement of warranties and insurance
xv	Appointment of consultants
xvi	Arrangement of the tender documentation
xvii	Selection of the procurement system
xviii	Organizing the contractor pre-qualification
xix	Participating in the evaluation of tenders and consequent selection and appointment of contractors
xx	Monitoring the progress
xxi	Organizing the control systems
xxii	Authorization of payments
xxiii	Arranging meetings
xxiv	Issuing safety health procedures
xxv	Organizing the communication and reporting systems
xxvi	Coordinating with the statutory authorities
xxvii	Addressing environmental aspects of the project
xxviii	Developing the final account
xxix	Monitoring the budget and variation orders
xxx	Organizing the handover
xxxi	Arranging the commissioning
xxxii	Organizing the maintenance manuals
xxxiii	Planning the maintenance programme
xxxiv	Planning for the maintenance period
xxxv	Arrangement for the feedback monitoring
xxxvi	Planning facility management

CPM is responsible for project's success. Few responsibilities of CPM as identified by Young, (2000) are:

- Identification and management of risks
- Selecting the core team with the project sponsors
- Project progress – tracking and monitoring
- Finding solutions to challenges that cause interference with project progress
- Taking responsibility and leadership initiatives for the project team

40 *Performance assessment of construction project managers*

- Completing the project deliverables along with benefits
- Performance assessment and management of the project team and all stakeholders involved

2.14 Value addition through construction project managers

Though the construction industry is experiencing large-scale investments; however, construction projects are consistently plagued by the issues of time and cost overruns. And the term "value" is a broad concept as nowadays all industries have moved to value-driven approaches which are governed by client-centric methodologies.

> In terms of construction projects: value can be defined as meeting the project/client requirements with minimum waste which can be achieved by application of project management processes in an effective manner.
>
> (Forgues, 2005)

As per Project Management Institute (2021), the application of construction management approach lies in both management and leadership activities.

> Construction project management can be defined as the implementation of skills, knowledge, tools, and techniques in order to achieve the objectives of construction of a built facility with the aim of ensuring completion of the project within the approved budget, scheduled time, and as per the quality standards.
>
> (Paul & Basu, 2021)

Good project management depends on the overall balance of time, cost, and quality in relation to building functionality and sustainability requirements. As the built environment is gathering great momentum, it's crucial to manage the construction projects effectively. The primary function of a CPM is the ability to add significant value to the overall process of project development throughout its lifecycle. This can be attained by the methodical application of a set of generic project-orientated management principles. The value that a CPM adds to a project is unique: no other procedures or activities can add comparable value, quantitative or qualitative, to any project (CIOB, 2014). The traditional approaches of project management are not sufficient in defining the functional and technical demands of a project and act as technical jargons that the stakeholders are not able to identify and relate them with their expectations that is where the role of a CPM comes into action and adds value by focusing on client requirements, with efficient streamlining of the processes of project management (Forgues, 2005).

The value added by a CPM to a project can be measured in terms of reduced cost, duration, risk, and better quality. As per ISO 9000, value addition at management level by a CPM is in terms of the following

Performance assessment of construction project managers 41

principles – leadership, customer focus, process approach establishment, engaging the people involved, relationship management, evidence-based decision making, and improvement in efficiency and productivity. CPMs are responsible for the successful delivery of a project that meets the organizational needs and adds value to the Client. They provide vision, direction, resolve issues, mitigate risks, and build better client relationships (Ali & Chileshe, 2009).

To summarize, as a team leader a CPM adds value to any project in the following ways (Ali & Chileshe, 2009; CIOB, 2014; IS 15883, 2015):

- Identification and specification of the roles and responsibilities of the team members
- Vision and objectives of the project
- Removal of impediments to a project
- Encouragement and development opportunities for the team members
- Effective assessment and management of risks
- Pro-active driver of a project
- Ensures a delighted client
- Ensures that the procedures are in place and are being followed
- Adds value in terms of value engineering exercise throughout the lifecycle of the project
- Performance monitoring
- Derives efficient construction methodology to be followed
- Optimizes every stage of the project
- Challenges the status quo

CPM acts as the most essential element responsible for project success. A major part of success is attributed to a CPM, as with growing transformations of technology the challenge for project manager is to keep one self-updated with the latest technology and implement the same for the entire project team (Bhangale & Devalkar, 2013).

2.15 Inferences

This chapter defines the roles and responsibilities of a project manager specific to the construction industry. Collecting the inferences from the above discussion, derived through the existing literature on project manager competence assessment, it clearly establishes the need to define a quantitative performance assessment measure of a CPM. The current body of knowledge of project manager competence assessment is based on the qualitative aspects of an individual; based on their personal competence. Though the standard by AIPM addresses technical competencies related to time, cost, and scope parameters but they are in a generic framework and lacks the major performance criteria for construction industry specific user requirements.

42 *Performance assessment of construction project managers*

In case of construction projects, the overall evaluation method for assessing the performance of a CPM is found to be less structured in a subjective manner. Hence, there is a need for establishing a more profound criterion for assessing a CPM's performance. The establishment of CPM's competency assessment framework would help in identifying the key competencies that are most likely to impact the performance of CPMs in the respective fields of the entire management process and are needed to be looked into further.

Based on the derived need for assessing CPM's performance with respect to project success requires a better understanding of the roles and responsibilities of a CPM, as the roles and responsibilities of a construction project manager are also a function of project typology.

The value addition through the assessment of CPM's performance lies in continuously upgrading and enhancing the level of skill set. As a CPM acts as the most essential element responsible for the final outcomes of a project.

References

Ahadzie, D. K., Proverbs, D. G., & Olomolaiye, P. O. (2008). Model for predicting the performance of project managers at the construction phase of mass house building projects. *Journal of Construction Engineering and Management, 134*(8). https://doi.org/10.1061/ASCE0733-93642008134:8618

Ali, M. M. A. A., & Chileshe, N. (2009). The influence of the project manager on the success of the construction project. *He 6th International Conference on Construction Project Management (ICCPM) | 3rd International Conference on Construction Engineering and Management (ICCEM) 'Global Convergence in Construction'ICCEM-ICCPM,* 1–24. https://www.researchgate.net/publication/265689336_The_influence_of_the_project_manager_on_the_success_of_the_construction_project

Atkinson, R. (1999). Project management: cost, time and quality, two best guesses and a phenomenon, its time to accept other success criteria. *International Journal of Project Management, 17*(6), 337–342. https://doi.org/10.1016/S0263-7863(98)00069-6

Bannerman, P. L. (2013). Barriers to project performance. *46th Hawaii International Conference on System Sciences,* 4324–4333. https://doi.org/10.1109/HICSS.2013.113

Bashir, R., Sajjad, A., Bashir, S., Latif, K. F., & Attiq, S. (2021). Project managers' competencies in international development projects: A Delphi Study. *SAGE Open, 11*(4). https://doi.org/10.1177/21582440211058188

Basu, T. (2015). Integrating 360 degree feedback in to performance appraisal tool and developmental process. *IOSR Journal of Business and Management (IOSR-JBM), 17*(1), 50–61.

Bhangale, P., & Devalkar, R. (2013). Study the importance of leadership in construction projects. *International Journal of Latest Trends in Engineering and Technology (IJLTET), 2*(3), 1–47.

Bohn, J. S., & Teizer, J. (2009). Benefits and barriers of monitoring construction activities using hi-resolution automated cameras. *Building a Sustainable*

Performance assessment of construction project managers 43

Future - Proceedings of the 2009 Construction Research Congress, 21–30. https:// doi.org/10.1061/41020(339)3

Bolzan De Rezende, L., & Blackwell, P. (2019). Project management competency framework. *Iberoamerican Journal of Project Management (IJoPM)*. *Www.Ijopm. Org*, *10*(1), 34–59. https://www.researchgate.net/publication/333882135

Callistus, T., & Clinton, A. (2016). Evaluating barriers to effective implementation of project monitoring and evaluation in the Ghanaian construction industry. *Procedia Engineering*, *164*, 389–394. https://doi.org/10.1016/J.PROENG.2016.11.635

Cartwright, C., & Yinger, M. (2007). *Project Manager Competency Development (PMCD) Framework.* PMI® Global Congress 2007—EMEA, Budapest, Hungary. Newtown Square, PA: Project Management Institute. https://www.pmi. org/learning/library/project-manager-competency-development-framework-7376

CIOB. (2014). *Code of practice for project management for construction and development.*- The charted institute of building. Fifth ed. Print ISBN: 9781118378083, Online ISBN: 9781118378168, https://doi.org/10.1002/9781118378168.

Cooke-Davies, T. J., & Arzymanow, A. (2003). The maturity of project management in different industries: An investigation into variations between project management models | Semantic Scholar. *International Journal of Project Management.* https://www.semanticscholar.org/paper/The-maturity-of-project-management-in-different-An-Cooke%E2%80%90Davies-Arzymanow/91fec3d430210f722df586 82d169135bcb0b591e

Demirkesen, S., & Ozorhon, B. (2017). Measuring project management performance: Case of construction industry. *EMJ - Engineering Management Journal*, *29*(4), 258–277. https://doi.org/10.1080/10429247.2017.1380579

Dombkins, D. H. (2007). *Complex project management.* https://books.google. co.in/books/about/Complex_Project_Management.html?id=j1_0IAAACAAJ& redir_esc=y

Forgues, D. (2005, September 13). A value-based approach to managing construction projects. *PMI® Global Congress.* https://www.pmi.org/learning/library/ value-based-approach-managing-construction-projects-7456

GAPPS. (2007). *A framework for performance based competency standards for global level 1 and 2 project managers.* www.globalpmstandards.orgsecretariat@ globalpmstandards.org

Hanna, A. S., Ibrahim, M. W., Lotfallah, W., Iskandar, K. A., & Russell, J. S. (2016). Modeling project manager competency: An integrated mathematical approach. *Journal of Construction Engineering and Management*, *142*(8), 04016029. https:// doi.org/10.1061/(asce)co.1943-7862.0001141

Hosain, Md. S. (2016). 360 Degree feedback as a technique of performance appraisal: Does it really work? *Asian Business Review*, *6*(1), 21. https://doi.org/10.18034/abr. v6i1.779

IS 15883. (2015). *Construction project management - guidelines: Part 1 general.*

Jones, J. E., & Bearley, W. 1. (1996). *360 Feedback: Strategies, tactics, and techniques for developing leaders* (M. George, Ed.). Human Resource Development Press. https://books.google.com/books/about/360_Degree_Feedback. html?id=k1_cNAnlMTYC

Kumar Das, U., & Panda, J. (2015). A literature review of 360-degree feedback as a tool of leadership development. *International Journal of Current Research*, *7*(4), 14757–14761.

44 Performance assessment of construction project managers

Lannon, J. (2019). Project management performance assessment in the non-profit sector. *Project Management Research and Practice*, 5. https://doi.org/10.5130/PMRP.V5I0.5910

London, M., & Smither, J. W. (1995). Can multi-source feedback change perceptions of goal accomplishment, self-evaluations, and performance-related outcomes? Theory-based applications and directions for research. *Personnel Psychology*, *48*(4), 803–839. https://doi.org/10.1111/J.1744-6570.1995.TB01782.X

Meshram, M., Gitty, R., Vinay, & Topkar, M. (2020). Project performance indicators for measuring construction performance in Mumbai. *International Journal of Engineering Research & Technology (IJERT)*, *9*(6). www.ijert.org

Mohammed, F., Al-Zwainy, S., Mohammed, I. A., & Raheem, S. H. (2016). Application project management methodology in construction sector: Review. *International Journal of Scientific & Engineering Research*, *7*(3). http://www.ijser.org

Nassar, N. K. (2009, October). An integrated framework for evaluation of performance of construction projects. *PMI® Global Congress*. https://www.pmi.org/learning/library/evaluation-performance-construction-projects-6751

Nijhuis, S., Vrijhoef, R., & Kessels, J. (2018). Tackling project management competence research. *Project Management Journal*, *49*(3), 62–81. https://doi.org/10.1177/8756972818770591

Oliveros, J., & Vaz-Serra, P. (2018). Construction project manager skills: A systematic literature review. *52nd International Conference of the Architectural Science Association*, pp. 185–192. The Architectural Science Association and RMIT. http://hdl.handle.net/11343/288752

Parker-Gore, S. (1996). Perception is reality: Using 360-degree appraisal against behavioural competences to effect organizational change and improve management performance. *Career Development International*, *1*(3), 24–27. https://doi.org/10.1108/13620439610118573/FULL/XML

Paul, V. K., & Basu, C. (2021). *A handbook for construction project planning and scheduling -*. COPAL Publishing Group. https://books.google.co.in/books?id=4eecDwAAQBAJ&printsec=frontcover&source=gbs_ge_summary_r&cad=0#v=onepage&q&f=false

Project Management Institute. (2017a). *A guide to project management body of knowledge (PMBOK guide): Vol. Sixth edition.*

Project Management Institute. (2017b). *Project manager competency development framework* (Vol. 3rd).

Project Management Institute. (2017c). *Project manager competency development framework: Vol. Third edition.*

Project Management Institute. (2021). *The standard for project management and a guide to the project management body of knowledge (PMBOK guide).* Seventh edition.

Rehman, A. U., Usmani, Y. S., & Al-Ahmari, A. M. A. (2012). A study to assess significance of project management performance: Assessment model in applied projects. *International Journal of Applied Systemic Studies*, *4*(3), 140–149. https://doi.org/10.1504/IJASS.2012.051129

Takim, R., & Akintoye, A. (2002). Performance indicators for successful construction project performance. In *University of Northumbria. Association of researchers in construction management* (Vol. 2).

Performance assessment of construction project managers   45

Toegel, G., & Conger, J. A. (2003). 360-degree assessment: Time for reinvention. *Academy of Management Learning & Education*, *2*(3), 297–311. https://doi.org/10.5465/amle.2003.10932156

Tornow, W. W., London, M., & Center for Creative Leadership. (1998). *Maximizing the value of 360-degree feedback: A process for successful individual and organizational development*. 291. Jossey-Bass.

Tunji-Olayeni, P., Mosaku, T. O., Fagbenle, O. I., Omuh, I. O., & Joshua, O. (2014). Evaluating construction project performance: A case of construction SMEs in Lagos, Nigeria. *Vision 2020: Sustainable Growth, Economic Development, and Global Competitiveness - Proceedings of the 23rd International Business Information Management Association Conference, IBIMA 2014*, *1*, 3081–3092. https://doi.org/10.5171/2016.482398

Udo, N. & Koppensteiner, S. (2004). What are the core competencies of a successful project manager? *Paper presented at PMI® Global Congress 2004—EMEA, Prague, Czech Republic, Newtown Square*, PA: Project Management Institute.

Unegbu, H. C. O., Yawas, D. S., & Dan-asabe, B. (2022). An investigation of the relationship between project performance measures and project management practices of construction projects for the construction industry in Nigeria. *Journal of King Saud University - Engineering Sciences*, *34*(4), 240–249. https://doi.org/10.1016/J.JKSUES.2020.10.001

Ustundag, A., & Cevikcan, E. (2018). *Industry 4.0: Managing the digital transformation*. https://doi.org/10.1007/978-3-319-57870-5

Wang, X. J., & H. J. (2016). The relationships between key stakeholders' project performance and project success perceptions of Chinese construction supervising engineers. *International Journal of Project Management*, *24*, 253–260. https://doi.org/https://doi.org/10.1016/j.ijproman.2005.11.006

Young, T. L. (2000). *Successful project management*. 155. https://books.google.com/books/about/Successful_Project_Management.html?id=JEzkqY_HODUC

Zwikael, O., & Globerson, S. (2006). Benchmarking of project planning and success in selected industries. *Benchmarking - an International Journal*, *13*(6), 688–700.

Zwikael, O., & Meredith, J. (2021). Evaluating the success of a project and the performance of its leaders. *IEEE Transactions on Engineering Management*, *68*(6), 1745–1757. https://doi.org/10.1109/TEM.2019.2925057

3 Performance index for construction project managers

3.1 Introduction

This chapter defines the concept of value drivers performance index (VDPI) for assessing any CPM's performance at individual as well as organizational levels. The process of deriving variables for defining VDPI has been covered in detail along with terminologies used for defining VDPI; the process of deriving variables for developing the performance index, quantification process, and the evaluation process for determining the VDPI value, the notion of multicollinearity in the developed VDPI equation indicating the interrelationship of the derived variables across the different stages of project lifecycle and its limitations have been described. Further, the chapter discusses about the application of VDPI at individual and organizational levels and its hardware user interface software application.

3.2 Concept of value drivers performance index (VDPI)

The concept of VDPI as represented in Figures 11 and 12 is developed using a comprehensive and state-of-the-art approach to index CPM's performance related to one's competence based on the current industry practices, requirements, and established codes and standards.

VDPI acts a performance assessment and enhancement tool for evaluation of CPMs that can be used by the organizations at different levels of the organizational hierarchy like individual level, individual-organizational level, multiportfolio organizational level, etc. There are certain non-tangible parameters which are needed to be evaluated as they act as a major source in value creation at both individual and organizational levels.

It identifies the key competencies of a project manager at different levels of hierarchy which impact the key determinants of any project's performance specific to the construction industry.

VDPI provides an individual/organization with one of the most precise results using a technologically structured interface derived from the extensive data processing and analysis and provides opportunities to determine the way of project handling techniques. The CPM skills and competency inevitably affect the quality of project execution. The managers of any

DOI: 10.1201/9781003322771-3

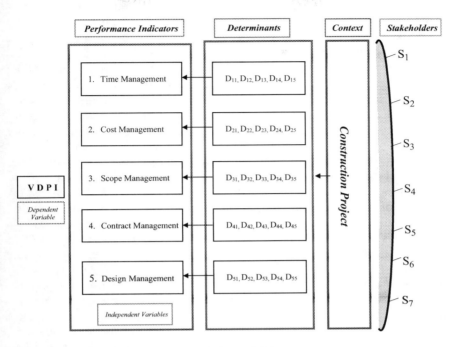

Figure 11 Value driver performance index (VDPI) concept.

project manager must be aware about how to assess the project manager's performance within their own organization and other organizations in the industry.

The major indicators used in deriving VDPI are based on construction project management professional's competencies/skills indicators and construction project performance indicators.

The VDPI is aimed at being indicative performance of the project manager measured over varied project attributes over the lifetime progression of the project professional. It is helpful for the CPMs and organizations with a direction to conduct assessment, planning, and management of the professional development of CPM along with the parameters of success.

The proposed VDPI would be calculated on the basis of scores obtained on various performance indicators which shall be moderated by corresponding weightages assigned to each indicator. The performance indicators themselves shall be derived on the basis of underlying attributes under each performance indicator which are determinants of performance.

3.3 Deriving variables for VDPI

The term variables used in VDPI refer to the key performance indicators for assessing the CPM's performance with respect to project success. In order

48 *Performance index for construction project managers*

to appropriately define success of any project, certain sets of performance variables need to be defined (Unegbu et al., 2022). The variables here defined are based on the skillset and knowledge application capability of a project manager specific to construction industry; they can be further defined in terms of organizations goals, project objectives which are based on the success parameters of a project and are considered to have some kind of impact on future endeavours of the individual or organization. However, the definition of success also varies for each individual. These variables used for deriving VDPI might vary from organization to organization based on different project scenarios, type/nature of organization, project typology, current industry trends and demands, etc. based on their parameters of defining performance and project success. While identifying these variables it is important to understand that these variables are to be used as a function for performance measurement of an individual or an organization which eventually helps in management decisions and in reducing the gap of individual's performance and industry benchmark. The variables might be impacted by the long-term and short-term goals of the project also based on their level of relevance. The selection of the variable inputs for assessing performance of a CPM might be different for all verticals within the organization and different projects. The variables must be accepted and appreciated within the organization. Hence, it is crucial to identify the variables in such a way that they provide a holistic overview of performance evaluation and should hold wider accountability. The identified variables should be such that they are not only measurable but also quantifiable (Lavy et al., 2014). Therefore, performance indicators should essentially cover all basic attributes of project management which can be measured and focus on critical aspects of outputs or outcomes. The process of deriving these variables might have different methodologies involving qualitative and quantitative approaches or a hybrid approach based on the organizational/individual purpose of measuring performance.

The variables that have been considered in the current chapter for developing the VDPI equation are gathered using hybrid approach of exploratory descriptive research using literature study, questionnaire survey, and expert interview.

Initially, an exhaustive list of variables for measuring construction project success is determined using the existing literature. An extensive amount of literature is available on the broader topic related to key variables affecting any project's performance in general which can be augmented to a concise list of key project performance variables. While the derived list of variables cannot be generalized for all construction projects, it can serve as a basis for the classification of project and organization-based performance measurement variables in developing the performance index and govern the progress road map.

The next stage involves expert discussions to finalize the key performance indicators governing project manager's performance with consideration to

Performance index for construction project managers   49

project success. On the basis of the review of existing literature and expert discussion, five broad categories of performance indicators have been considered for determining the VDPI score for evaluating CPM's performance are time, cost, scope, design, and contract based on the management areas of Project Management Body of Knowledge (PMBoK) guide. The identified performance variables for assessing the performance of construction managers are later termed as performance indicators. As mentioned in Chapter 2, the key performance indicators are generally referred as project performance indicators of time, cost, scope, design, and contract.

Further, a questionnaire survey was conducted to derive the weightages of each performance indicator for measuring project manager's performance. Initially, a pilot study was conducted among a group of 20 participants belonging to the construction project management background in Indian industry domain, to modify any discrepancies.

The aim of the questionnaire survey was to derive the weightages of the identified variables for assessing the CPM's performance in the industry based on the processes of project management as listed below:

- Significance of time, cost, scope, contract, and design management performance for assessing CPM's performance.
- Significance of determinants of time management performance:
  a   Planning work coordination
  b   Effective schedule control
  c   Risk forecasting
  d   Effective resource planning
- Significance of determinants of cost management performance:
  a   Effective cashflow management
  b   Controlling budget variance
  c   Managing risk contingencies
  d   Controlling cost overruns
- Significance of determinants of scope management performance:
  a   Coordinated scope planning
  b   Effective stakeholder involvement
  c   Monitoring project deliverables
  d   Controlling scope creep
- Significance of determinants of contract management performance:
  a   Risk-sensitive procurement planning
  b   Effective planning of contractual obligations
  c   Effective management of contractual obligations
  d   Effective claim management
  e   Planning contract closeout
- Significance of determinants of design management performance:
  a   Establishing stakeholder engagement process
  b   Establishing needs centric design process
  c   Establishing decision-making hierarchy

50 *Performance index for construction project managers*

d   Resolving conflicting interests
e   Effective planning for scope creep
f   Resolving time-cost impacts

The identified performance indicators are broad areas for defining project success. It is better to break down the indicators into determinants for measuring the performance of each indicator at the ease of an individual assessor. The variables which are considered for determining any project performance variables are termed as their determinants in the VDPI concept. The determinants of each performance indicator have been identified through in-depth literature review and have been revised and modified through expert interviews. Further, these determinants are mapped along the knowledge areas of project management as defined in PMBoK guide.

The next section comprising of each performance indicator and their respective determinants has been structured based on a 5-point Likert scale so that participants chose the importance rating most closely aligned to their own experience for each question. The participants were asked to rate each determinant on a 5-point Likert scale, where 1 – Least Important, 2 – Somewhat Important, 3 – Neutral, 4 – Important, and 5 – Very Important.

The final responses were obtained and weightages for each determinant were derived which are further used in Section 3.4 of the current chapter for defining the quantification process of VDPI.

Further, the survey analysis results are used for determining the weightage of the performance indicators as well as the determinants of each performance indicator which will be used in the calculation of the VDPI for any CPM and are used in the illustration of assessing the performance of a CPM in Chapter 10.

Based on the understanding of the project individuals involved in the project and as per the organizational context, objectives, and the convergence of the derived variables, organization-specific approach based on the typology of projects and organizational requirements can be carried out to derive the weightages of the performance variables and their determinants, so that any organization applying the concept of deriving variables and weightages for assessing VDPI of an individual CPM or at the other levels of hierarchy can formulate a questionnaire survey or use any other method to derive these weightages which will eventually be governed by both the organizational and project specific requirements.

3.4 Quantification process

The methodology for calculation of VDPI is based on majorly two concepts of performance indicators – project performance (time, cost, scope, design, quality, etc.) and CPM's competencies/skill performance indicators (business acumen, technical skills, ethics, etc.).

Performance index for construction project managers 51

The performance here refers to what a person is able to execute and achieve by applying their knowledge and skills. As described in PMBoK, the determination of performance can be in terms of the delivery of successful projects and qualification of credentials. The process of VDPI calculation is represented in Figure 12.

$$\text{VDPI} = \sum f\left(P_n, W_n, C_k\right) \qquad \qquad \textit{Equation 1}$$

As can be observed from the above expression, VDPI is a function of P, the Performance Indicator Score, its weightage W, and a variable C which measures the collinearity between various factors as well as determinants to account for the interrelationships between the factors within a group or among different groups. It would be reasonable to assume linearity between the indicators as well as determinants among themselves. This assumption is made on the basis of the reasoning that the correlations within as well as among groups would not cause a significant change in the overall evaluation. However, if an evaluator wishes to refine the evaluation and account for such interactions, the necessary values of C can be evaluated using appropriate techniques and equation of the VDPI can be suitably modified by introducing factors that would account for this variation among as well as within the groups.

The VDPI shall be calculated as a weighted summation of scores achieved under different performance indicators and shall be expressed as

$$P_n = \sum_{i=0}^{m} D_m \text{x } W_m \qquad \qquad \textit{Equation 2}$$

Assuming effect of intercorrelations as negligible, the equation for VDPI can be suitably modified as follows:

$$\text{VDPI} = \sum_{i=0}^{n} P_n \text{ x } W_n \text{ x} Ck \qquad \qquad \textit{Equation 3}$$

where

- P_n is the nth performance indicator which would vary from organization to organization but would include basic performance indicators measured in respect of time, cost, scope contract, and design management. However, these indicators can be set by any organization as per their own methodology and requirements.
- D_m is the determinant score of each determinant of performance indicator on a scale of 1 to 10.
- W_m is the weightage accorded to each performance determinant which shall be calculated basis expert ratings on relative importance of the determinants of the performance under each indicator.

52 *Performance index for construction project managers*

- W_n is the weightage of each performance indicator calculated on the basis of experts' relative importance of the indicators in overall evaluation of VDPI.
- *Ck* is mutual interference constant.
- Value of *Ck* is always greater than 0.
- *Ck* < 1, if P1, P2,...,Pn have strong negative mutual interference as determined in construction project context.
- *Ck* = 1, if P1, P2,...,Pn are unique and do not interfere with each other.
- *Ck* > 1, if P1, P2,...,Pn have strong positive mutual interference as determined in construction project context.

The above expression is a generic expression that can be suitably modified by any organization or individual to suit their requirement.

The ease of application of VDPI lies in development of a cloud-based tool, which can be easily accessed by its users. The theoretical understanding on pen and paper can be used as a one-time exercise, but considering the size of an organization and limitation of the scale of data, a cloud-based integrated data processing system can be more useful. The premise is that there is continuous performance assessment and improvement as envisaged in total quality management approach for the customer (internal/external).

3.4.1 Performance indicators and their determinants

Performance measures in the context of construction project management are the numerical or quantitative indicators of the performance. Therefore, the indicators provide measurable evidences required to showcase that the planned techniques are valuable as desired outcomes have been achieved. However, when it is not possible to obtain a precise measurement, performance indicators are usually referred. For performance measurement to be effective, the measures or indicators must be accepted, understood, and owned across the organization. Therefore, performance indicators should essentially cover all basic attributes of project management which can be measured and focus on critical aspects of outputs or outcomes.

These performance indicators (PIs-P1, P2,...,Pn) are derived with respect to construction management at both individual and organizational levels. The PIs consist of some of the most important performance objectives across all aspects of an individual/ organization involvement at project and organizational levels. Accordingly, indicators of performance have been summarized from literature and crystallized into five major performance indicators, as noted in Table 4, through review, questionnaire survey, and discussions with experts. These indicators encompass every aspect of project management and hence would qualify as being representative of all expected outcomes of a project manager's performance.

The terms and range of an organization's PIs tend to differ for each project for different verticals of an organization. Hence, the performance indicators

Performance index for construction project managers 53

are identified in such a way that they essentially cover all basic attributes of project management which can be measured and focus on critical aspects of outputs or outcomes. The relative importance of each performance indicator and its determinants might vary with respect to an individual as well as organization. The performance indicators may need to evolve and change depending on the methodologies adapted by various organizations. Accordingly, the number of indicators and their terminologies can be varied as per the needs and requirements of the evaluator who can be an individual or an organization.

The competencies are too broad to be measured and quantified directly, so the determinants of individual indicators are used to derive the weightages for each performance indicator.

The underlying value (P) of each indicator is derived from various determinants of that particular indicator, identified again from the literature. These determinants would have to be scored for each project manager, either by himself as a part of self-assessment or by the organization as a part of project performance evaluation on a scale of 1 to 10, 10 being the best, 01 being the worst or any other scale deemed fit by the evaluator.

The weightage of these performance indicators is derived using various determinants (D1, D2,...,Dn). A survey of experts is used to rank the determinants of each performance indicator in order of their importance and a relative ranking scale is used to determine the relative weightages of each determinant. Thus, a relative score for each performance indicator is calculated and VDPI is then the summation of all individual performance indicator scores.

These determinants are needed to be scored on a standard scoring scale as a part of self-assessment by an individual or as a part of an individual organizational project performance evaluation, or as a part of multiportfolio organizational level evaluation. These determinants are also further assigned a weighted score (w11, w12,..., wm), which ultimately provides the customer/user with their VDPI score (construction project manager performance score).

Researchers and academicians have reported that the traditional approach for measuring project performance is just based on time, cost, and quality, which can be crude for evaluation of any project manager's performance. Given the complex set of variables that affect project performance, it is advisable to differentiate between commonly used vs which should be used parameters. The identification of determinants of the performance indicators is carried out through extensive literature review.

VDPI is then calculated as the summation of all individual performance indicator scores as mentioned in $P_n = \sum_{i=0}^{m} D_m \times W_m$ Equation 2. Based on the established benchmarks, the scores can be evaluated at both individual and organizational levels.

Table 4 A summarized view of the indicators and their determinants is produced in the form of a matrix

S.no	Performance indicators	Weightage of indicators	Determinants of performance		
	(Pn)	(Wn)	m	Wm	(Dm)
1	Time performance	W1	1	W11	Planning work coordination
			2	W12	Effective schedule control
			3	W13	Competence for Resource planning and control
			4	W14	Risk management approach
2	Cost performance	W2	1	W21	Effective cash flow management
			2	W22	Controlling budget variance
			3	W23	Managing risk contingencies
			4	W24	Controlling cost overruns
3	Scope performance	W3	1	W31	Coordinating scope planning
			2	W32	Effective stakeholder involvement
			3	W33	Monitoring project deliverables
			4	W34	Controlling scope creep
4	Contract	W4	1	W41	Risk-sensitive procurement planning
			2	W42	Effective planning of contractual obligations
			3	W43	Effective management of contractual obligations
			4	W44	Effective claim management
			5	W45	Planning contract closeout
5	Design performance	W5	1	W51	Establishing stakeholder engagement process
			2	W52	Establishing need centric design process
			3	W53	Establishing decision making hierarchy
			4	W54	Resolving conflicting interests
			5	W55	Effective planning for scope creep
			6	W55	Resolving time-cost impacts

Performance index for construction project managers   55

The final evaluation results can be generated in the form of reports which can be interpreted by the users, based on the information as needed from various other perspectives.

3.5 Evaluation process of VDPI

The overall evaluation process of VDPI is very simple and straight forward, offering a quantitative assessment approach, involving a set of performance measurement indicators and their determinants derived in a very consistent manner and scored by the individuals.

There are two ways of evaluating the performance by using VDPI:

- Self-appraisal
- Organizational appraisal

During the self-appraisal evaluation process, an individual rates his/her performance with respect to the robust list of available determinants against each performance indicator with regard to their individual projects context and gets a final score S1. The self-appraisal process is for the evaluation of individual input units (IUs) and project hosts (PHs, individual project managers) at individual levels, which requires the input in the form of project field inputs and based on the analysis of the self-appraisal scores. Through the calculated VDPI scores, it is decided whether the performance is below, equal to, or above the required standards and aids in the identification of major gap areas requiring further improvement. Accordingly, the goals can be set and development strategy can be formulated as per the evaluated VDPI score. The scores can be evaluated later, based on the industrial identified benchmarks by the individual. One of the major advantages of VDPI is that these performance indicators can be customized as per the individual's roles on the project/organization.

In the case of organizational appraisal evaluation process, the performance of the organization is measured. This measure of organizational performance is for all verticals being operated by the organization separately (Organizational unit, OU) and it also offers complete organizations multi-portfolio performance assessment (Orgn unit). The details for the input and process can be referred in the next chapter of this book focusing on the technical support device and its hardware interface.

The evaluation process at the organizational level is the same as that in self-appraisal evaluation process. Here, the indicators are to be scored considering the organizational perspective. The input data in case of organization unit are the results of individual outputs of the self-appraisal process of project hosts which are used in the appraisal process of the organization and the final VDPI score of the organization O1 is obtained based on the industrial benchmarking to determine the organization's level of performance. Just as it is in the case of self-appraisal, the obtained score, O1 is used to

56 *Performance index for construction project managers*

compare the final organizational performance. The final scores provide insights into the major leading and lagging indicators which represent the need for improvement.

These scores obtained can also be used to assess the individual's performance with respect to organizational requirements.

- $O1-S1>0$, Need for improvement
- $O1-S1=0$, Individual meets organizational requirements
- $O1-S1<0$, Individual exceeds organizational requirements

If any individual/organization is unsatisfied with their VDPI score, they can re-evaluate their scores subject to improvement in their competencies. The cross-business strategies can also be formulated into the performance indicators to assess the overall improvement in the performance of an organization.

The obtained VDPI score helps in better decision-making at both individual and organizational levels. While VDPI provides an overall performance evaluation of a CPM, calculating individual performance indicator scores enables one to gauge an individual's improvement areas under different PM spheres, that when acted upon would provide the required improvement to drive value in the overall project management sphere.

VDPI is a flexible, effective, and user-friendly performance measurement tool for the evaluation of CPM's performance using a quantitative approach, based on project performance criteria. It acts as a support tool for a CPM, for assessing the competencies of a PM and comparing their performance to the industry and set benchmarks for better evaluation in the next assessment, leading towards a more balanced, well-equipped method for evaluation of the performance of an individual CPM or at an organizational level. The applicability of VDPI lies in its use in self-assessment by the user during all stages of the construction project's lifecycle.

The main purpose of VDPI is to help individuals as well as organizations in transforming their goals into tangible performance indicators, which helps them in evaluating the areas in need for improvement. The VDPI should not be considered as a benchmarking exercise.

3.5.1 *Multicollinearity of performance indicators*

Multicollinearity is a statistical phenomenon and it generally occurs in the case of linear regression analysis when two or more variables are found to be mutually correlated and do not provide any unique or independent information.

It may be noted here that multicollinearity presented below is one such tool that appears to be most suited to address the problem of VDPI across stages that are simultaneous, concurrent, and interdependent. However,

Performance index for construction project managers 57

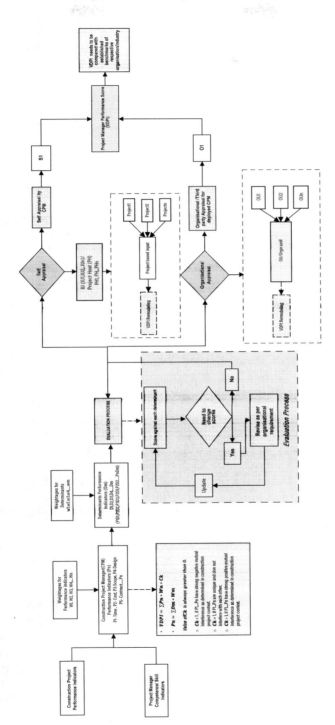

Figure 12 VDPI process methodology flowchart.

58   *Performance index for construction project managers*

there may be other techniques suitable; it is therefore expected that the reader explores other techniques to identify the most suitable one.

Multicollinearity, also called near-linear dependence, is a statistical phenomenon in which two or more predictors or independent variables as they are called in a multiple regression model are highly correlated. If there is no linear relationship between predictor variables, they are said to be orthogonal (Jensen & Ramirez, 2012). The term multicollinearity was introduced by Ragnar Frisch, meaning that there is a perfect relationship among some or all explanatory variables (Alin, 2010).

The aim of any regression model is to isolate the relationship between each independent variable and dependent variable. The coefficients obtained through analysis are in fact nothing but measures by which dependent variables change with respect to a unit change in independent variables, keeping all other independent variables constant. However, if correlations are present among independent variables, and if the value of any independent variable is changed, the one correlated with this variable will also exhibit some change. Hence, the stronger this relationship among variables, the more it becomes difficult to keep the other independents constant and change only one independent variable. Consequently, it becomes difficult for the model to distinguish between the effect of one variable and the other or to put it differently, it becomes difficult for the model to estimate the relationship between each independent variable and the dependent variable independently because the independent variables tend to change in together (Siegel & Wagner, 2022). This is called the problem of multicollinearity. Whenever two supposedly independent variables are highly correlated, it will be difficult to assess their relative importance in determining some dependent variable. The higher the correlation between independent variables the greater the sampling error (Blalock H.M, 1963).

The basic assumption in any linear regression analysis is that there is no multicollinearity among independent variables. Researchers (Gujarati & Porter, 2003) in their book argue that the reason for this is that if multicollinearity is perfect among any independent variables, the regression coefficients of independent variables are indeterminate and their standard errors are infinite. If the multicollinearity is not perfect, that is, it is not very high (perfect means R value of 1) the regression coefficients, although determinate, have large standard errors, which means the coefficients cannot be estimated with great precision or accuracy (Gujarati & Porter, 2003).

Multicollinearity can have two effects, and therefore, is said to be of two main types viz Statistical and Numerical. Statistical consequence is concerned with difficulties in ascertaining and testing individual regression coefficients due to high standard errors leading one to declare an independent variable as less significant, though it might be highly related to dependent variable. Numerical consequence on the other hand relates to difficulties in numerical calculations in the software tools due to instability in coefficients.

Performance index for construction project managers 59

This can lead to either failed analysis or reporting of meaningless values by the program.

However, researchers have argued that multicollinearity may or may not be a problem as it not only depends upon the extent or strength of correlations but also on the purpose of analysis (Siegel & Wagner, 2022). If the purpose of the study is primarily to predict or forecast a dependent variable Y (Which is P as well as VDPI in our case), strong multicollinearity may not be a problem because a careful multiple regression program can still produce the best (least squares) forecasts of Y based on all of the independent variables X (Which are Determinants as well as PIs in our case). However, if one wants to use the individual regression coefficients to explain how Y is affected by each X variable, then the statistical consequences of multicollinearity will be significant because these effects cannot be separated (Siegel & Wagner, 2022).

Therefore, the fact that VDPI is only trying to estimate the overall performance value, the statistical significance of multicollinearity can be neglected. Moreover, researchers have also argued that the problem of multicollinearity should not be viewed in isolation and that a high value of R^2 and a large sample size can offset the problems caused by multicollinearity (Jensen & Ramirez, 2012) which is again confirmed by Grewal et al. (2004) in their study. This points to the fact that any user or organization intending to use VDPI must be careful in securing enough sample size for ascertaining the values of determinants so that the effect of multicollinearity, if any, is minimized.

Let us try to analyze the equation of VDPI. The equation of VDPI illustrates that it is a function of performance indicator score (P_n) and their weightages (W_n).

$$\text{VDPI} = \sum f(P_n, W_n, C_k)$$

The presence of factor C_k in the above expression is only indicating that the expression is, besides being a function of P, a function of multicollinearity between variables, that is, between determinants within a particular performance indicator, between determinants of two or more performance indicators, as well as correlation, if any, between performance indicators themselves. This essentially means that there is probability that the determinants (variables) can have some degree of correlation between them and even the performance indicators could have some correlation among themselves. Therefore, the presence of factor C_k in above expression is to only reiterate the fact that correlation between variables for the purpose of illustrating the concept of VDPI and readers or the users are encouraged to test their variables for multicollinearity and use appropriate methods for making the model. Theoretically, if the expression of the VDPI was to be presented pictorially, it would look like something presented in Figure 13.

60 Performance index for construction project managers

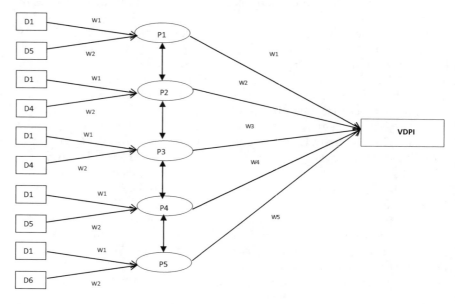

Figure 13 Correlation between the variables.

In Figure 13, the curved arrows between determinants indicate correlation between the determinants while the straight arrows between performance indicators represent correlation between the PI themselves. The calculation of VDPI may be modified according to the strength of these correlations, though, we believe that such correlations can be neglected for the purpose of the VDPI as the aim of VDPI is to only predict the value of performance from determinants. We will come to this point later in the chapter.

The equation for the calculation of VDPI is developed using the multiple linear regression model, where the dependent variable is the VDPI value and the independent/exploratory/explanatory/predictor variables are time, cost, scope, design, and contract performance indicators.

$$Y = a + b*X1 + c*X2 + d*X3 + e*X4 + f*X5 + \epsilon \qquad \text{Equation 4}$$

where Y: Value drivers performance index (Dependent variable)
 X1: Time performance management indicator
 X2: Cost performance management indicator
 X3: Scope performance management indicator
 X4: Design performance management indicator
 X5: Contract performance management indicator
 b, c, d, e, f: slopes
 a: intercept
 ε: residual error

Performance index for construction project managers 61

Therefore, if we discount the effect of multicollinearity, the calculation of VDPI can be reduced to a linear summation of the product of performance indicator score and the weightages and the equation becomes -

$$VDPI = \sum_{i=0}^{n} P_n \; x \; W_n$$

where performance indicator score P_n is derived from determinants score as given below

$$P_n = \sum_{i=0}^{m} D_m x \; W_m$$

The assumption that C can be neglected so that the equation of VDPI remains linear in addition of only PI Scores is akin to neglect the effect of multicollinearity, as explained above, between and among the variables. The discounting of multicollinearity from the expression will need some explanation which is provided in the following paragraphs. However, it is reiterated that organizations or users may model the VDPI to account for such correlations by using appropriate techniques if they believe that the intercorrelations are too strong among the variables.

3.5.1.1 Dealing with multicollinearity: other tools

From the above discussion, it can be observed that the problem of multi-collinearity can be mitigated by using a large sample size, which in effect means using a strong and large database. This has been discussed later in the book. Researchers have also suggested using other methods of analysis to mitigate this issue and one of the suggested methods is that of structural equation modelling or SEM. Many researchers seem to think that structural equation models are robust against multicollinearity (Malhotra et al., 1999) while others have opined that using SEM can provide necessary remedy in multicollinearity problems (Verbeke & Bagozzi, 2000) and can help "deal with some cases where the correlations among predictors are high" (Maruyama, 2014). One of the reasons put forward for this belief is that if highly correlated variables can be regarded as indicators of a common underlying construct, multicollinearity problems can be avoided (Grewal et al., 2004).

Therefore, organizations or individuals wanting to use VDPI can model the VDPI equation using such techniques like SEM to minimize, if not completely remove the effect of multicollinearity. Also, another point that shall be kept in mind while modelling VDPI is that though methods like SEM are believed to have a remedial effect on collinearity, the process adopted should, nevertheless be such that it is capable of checking the extent of the strength of these correlations and procedures may be adopted to reduce the

62 *Performance index for construction project managers*

correlations to the extent possible. Numerous methods are prescribed in literature for doing so.

3.5.2 *Relationship of variables across project lifecycle*

The variables used in defining the VDPI equation, for assessing the CPM's performance, are majorly related to each other and are found to be acting as independent and dependent variables throughout the series of a project.

The identified variables for the assessment of project's scope, time, cost, design, and contract management performance are found to be significantly related with the project performance parameters. We are well aware of the different stages of any project's lifecycle, as defined in Project Management Institute (2017)

- Project Initiation Phase
- Project Planning Phase
- Project Execution Phase
- Project Monitoring and Control Phase
- Project Closure Phase

Each stage in any project's lifecycle has a different focus related to project performance. Project lifecycles can be predictive and adaptive, sometimes the phases can be sequential, iterative, or overlapping within a project lifecycle. Predictive life cycle is also termed as waterfall. In such cases, project time, cost, and scope are determined in the early phases of the project. In case of iterative project life cycle, the scope of the project is only determined in the early stage and later on time and cost can be modified based on project understanding and further developments (Project Management Institute, 2017).

Project scope, time, cost, design, and contract performance parameters are always mutually connected to one another and define the success criteria for any project.

The challenge of relating VDPI and applying it in the continuum of project stages is presented in Figure 14.

The complex relationship between the variables is illustrated in Figure 14.

The complexity of performance of construction project manager within the continuum of project across the stages and the implications thereof are presented in Table 5.

The way forward requires to resolve this complexity through a mathematical equation involving multicollinearity or an equivalent method which in current VDPI equation is termed as C_k to give subjective multiplier and the wisdom of CPM (their learnings and understanding). The multicollinearity is explained in detail in the subsequent section.

Performance index for construction project managers 63

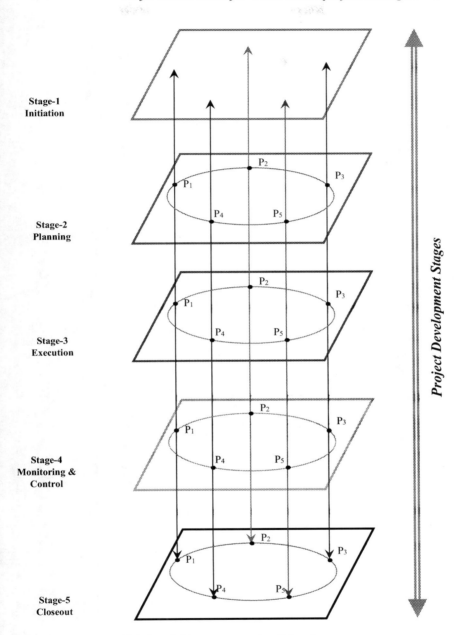

Figure 14 Relationship of variables across project lifecycle.

Table 5 Relationship of time performance variables across project lifecycle

Performance indicators	Weightage of indicators	Determinants of performance (identified from literature, expert interviews, and field studies; weightages (Wm)			Relationships of variables (performance indicators)				
(Pn)	(Wn)	m	Wm	Determinants	Initiation	Planning	Bid and award	Execution	Commissioning and closeout
Time Performance	W1	1	W11	Planning work coordination					
		2	W12	Effective schedule control					
		3	W13	Risk forecasting					
		4	W14	Effective resource planning					
		5	W15	Controlling delays					

Table 6 Relationship of cost performance variables across project lifecycle

Performance indicators	Weightage of indicators	Determinants of performance (identified from literature, expert interviews, and field studies); weightages (Wm)			Relationships of variables (performance indicators)				
(Pn)	(Wn)	m	Wm	Determinants	Initiation	Planning	Bid and award	Execution	Commissioning and closeout
Cost Performance	W2	1	W21	Effective cash flow management					
		2	W22	Controlling budget variance					
		3	W23	Managing risk contingencies					
		4	W24	Controlling cost overruns					

Table 7 Relationship of scope performance variables across project lifecycle

Performance indicators	Weightage of indicators	Determinants of performance (identified from literature, expert interviews and field studies); weightages (Wm)		Relationships of variables (performance indicators)				
(Pn)	(Wn)	m	Wm	Determinants	Initiation	Planning	Bid and award	Execution	Commissioning and closeout
Scope Performance	W3	1	W31	Coordinating scope planning					
		2	W32	Effective stakeholder involvement					
		3	W33	Monitoring project deliverables					
		4	W34	Controlling scope creep					

Table 8 Relationship of contract performance variables across project lifecycle

Performance indicators	Weightage of indication	Determinants of Performance (identified from literature, expert interviews and field studies); weightages (W_m)			Relationships of variables (performance indicators)				
(Pn)	(Wn)	m	Wm	Determinants	Initiation	Planning	Bid and award	Execution	Commissioning and closeout
Contract	W4	1	W41	Risk-sensitive procurement planning					
		2	W42	Planning contractual obligations					
		3	W43	Managing contractual obligations					
		4	W44	Effective claim management					
		5	W45	Planning contract closeout					

Table 9 Relationship of design performance variables across project lifecycle

Performance indicators	Weightage of indicators (equal)	Determinants of performance (identified from literature, expert interviews and field studies); weightages (W_m)			Relationships of variables (performance indicators)				
(Pn)	(Wn)	m	Wm	Determinants	Initiation	Planning	Bid and award	Execution	Commissioning and closeout
		1	W51	Establishing stakeholder engagement processes					
		2	W52	Establishing design process					
Design performance	W5	3	W53	Establishing decision making hierarchy					
		4	W54	Resolving conflicting interests					
		5	W55	Effective planning for scope creep					
		6	W56	Resolving time-cost impacts					

Performance index for construction project managers

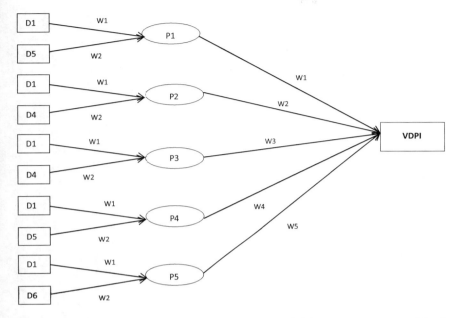

Figure 15 Correlations between the independent variables are not accounted for in existing model.

3.5.3 *Limitations*

As mentioned in the above paragraphs, the effect of multicollinearity might have a limiting effect on the reliability of the construct, but such a limitation may not necessarily manifest in the modelling for the reasons discussed above. One of the ways of reducing the effect, as already mentioned above, is to use a large sample space. In case of organizations, this would essentially mean to have a large database on determinants and performance indicators. Therefore, before trying to implement the VDPI and model it to suit the requirements of a particular organization, it is a prerequisite for the organizations to develop organizational process maturity through which they can build databases suitable and sufficient enough to model VDPI using appropriate techniques, so that the effect of intercorrelations do not have a significant impact on the overall index. For such organizations, the VDPI model may then look like the one provided in Figure 15, where correlations between the independent variables are not accounted for.

Such a model then easily fits the linear equation model mentioned earlier. The quality of the determinants and hence the performance indicator score will keep on improving with continuous enhancement and maturity in the knowledge database of the organization which can be only ensured through a robust organizational process approach.

70 *Performance index for construction project managers*

3.6 Application of VDPI

The wider applicability of VDPI tool is within the construction industry domain focusing on the project management perspective. The major contribution of VDPI lies in the processing of performance assessment indicators based on a comprehensive set of competencies. It may act as an important step towards better devaluing the complexities and nuances of a construction management professional.

The VDPI tool can be developed into an app-based (android/IOS) software system considering its application in the construction industry that can be easily accessed on hardware devices like mobile phones, tabs, laptops, etc. The hardware interface for the supporting device is for the field inputs only which can be used by anyone at project sites for uploading of relevant data on the application platform, which act as input nodes for the customers/users for entering the data (scoring of the performance indicators at individual level). The hardware interface is for input only, and the remaining mechanism is completely software-based design.

Since a manual data entry might turn out to be a very time-consuming exercise, the help of standard software system and the use of VDPI by each individual becomes quite approachable, considering the nature of the construction industry in terms of project-specific locations. The process of data entry and processing is highly confidential, so as to make sure that the entire assessment turns out to be true in terms of the scores allotted by the user and is only accessible to the authorized designated user.

In the proposed support device, there are four hardware interfaces coined in the VDPI interface:

i   Input unit (IU)

The input unit is basically an individual level (project coordinator/assistant) that feeds the required input data in the software interface by using any supporting hardware device (hand held, potable, independent).

This deals with the assigning of scores at individual levels which is dependent on the project field data input. The details of the hardware interface for IU are presented in Figure 16.

The data needed to input is related to an individual's roles and responsibilities as defined in the project charter related to the individual project entity like design management, commercial management, contract management, etc. The personal competency of an individual can be described by indicators such as influence, communication, contextual management professionalism, knowledge, experience, leadership, etc. These input parameters are scored and analyzed with final outputs of achieved VDPI scores.

Further, the collected information is processed to the project host (project manager). An individual VDPI score is generated which can be interpreted based on an individual's self-assessment.

Performance index for construction project managers

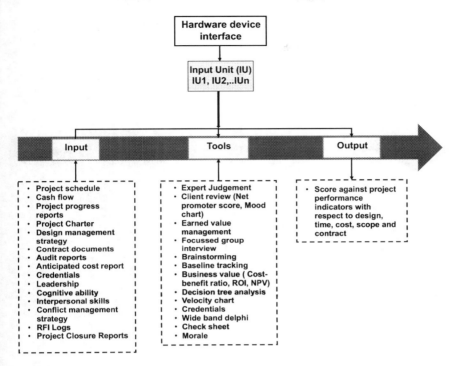

Figure 16 Hardware device processing phase process for IU.

In Figure 2, the input units are represented as IU1, IU2, ..., IUn whose final outputs are saved in the main server.

ii Project host (PH)

The project host can be the assistant project manager or project manager. Sometimes there can be more than one project managers in a project depending upon the size of the project. Construction industry projects typically involve safety manager, quality manager, execution manager, planning manager, design manager, commercial manager, etc., based on the scope of work; they report to a senior manager handling the project. Such senior managers generally are responsible for handling of multiple projects.

The processing of PH is similar to that of IU; the project host works on the fixed device software interface. The project manager might be responsible for multiple projects at the same time, so the project host might be evaluating the performance based on the data of multiple projects and obtaining a VDPI score, which can be further examined as per the suitability of techniques available like CPM assessment on project, CPM benchmarking on project as represented in Figure 17.

72 Performance index for construction project managers

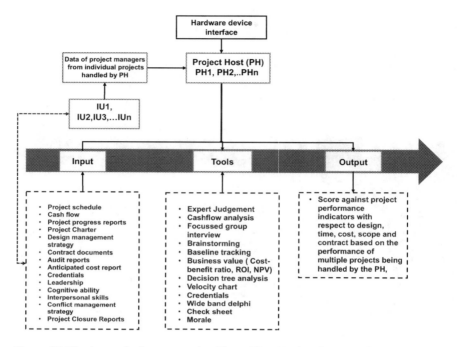

Figure 17 Hardware device processing Phase Plan-Project Host (PH)

The CPM's performance can be evaluated and compared using the benchmarking model, which would cater towards continuous improvement of the CPM and would help in the formulation of the development strategy. For the android/IOS app-based interface the CPM is defined as the project host. The project hosts are represented by PH1, PH2,..., PHn in Figure 4, whose data is also stored in the main server, and based on the organizational requirements, the final outputs are also used as an input variable for organization unit.

iii Organization unit (OU)

The organizational unit in the current device support system is the data which is being used at the organization level as an individual organization operates in various verticals.

At OU interface, the data from various project hosts in the respective vertical is collaborated for all projects' performance within the respective vertical of the organization, for instance, interior fit-outs, and it is entered as an input to the OU unit. Further, the data analysis can be carried out using established organizational tools of analysis. At OU level, it is majorly the payback period, net present value, and ROI (return on investment).

Performance index for construction project managers

Hence, an organization can input data based on its different operating verticals and can self-assess its performance using the VDPI score criteria. These scores can be defined based on each individual project's performance parameters and competency-related indicators of as derived from the outputs generated through the project hosts of different projects within the organization.

In the present model of VDPI, the organization is defined in terms of all projects belonging to a single vertical within the organization like interior fit out, highways, airports, mixed-use development projects, etc. within each vertical, and there will be multiple projects associated with each individual vertical division being handled by individual project managers (PH) who will be sharing their individual VDPI scores. At the organizational level, the major input parameters will be in terms of net business value generated, such as net present value, payback period, etc. Based on the achieved VDPI scores, the same can be analyzed and compared with other verticals within the organization. The organizational VDPI score represents the use of organizational-level self-assessment and organizational-level benchmarking. For diagrammatic representation please refer to Figure 18.

Many organizations also use assessment and provide feedback on the performance of an individual. However, these existing methods don't look at the individual performance from the project performance point of view.

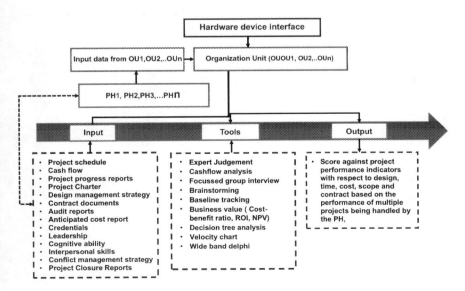

Figure 18 Hardware device processing Phase Plan-Organization Unit (OU)

iv   Multiportfolio CPM (Orgn unit)
   In the case of Multiportfolio CPM (Orgn unit), there are multiple verticals within the organization, in which respective goods and services are provided. In the construction industry domain, these services are majorly related to develop, consult, operate, construct, etc., for different categories of projects like airports, highways, waterways, high-rise residential townships, commercial projects, mixed-use developments, etc. In case of VDPI evaluation of multiportfolio process, it encompasses the overall performance of an organization. The inputs for Orgn unit are the results received from OU units combined together into evaluation dashboards and trackers, which help an organization in accessing the performance.
   The input for Orgn unit is mainly the combination of outputs as received from the other hardware interfaces which are critically analyzed from multi-level organizational structure decision tree support system. The Orgn unit interface hardware process is presented in Figure 19.
   The complete data along with the analysis results is stored in the main server of the VDPI support device.

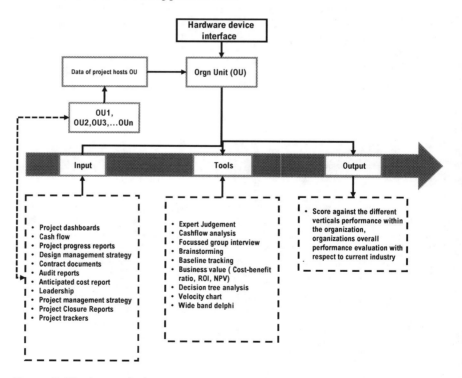

Figure 19 Hardware device processing Phase Plan-Multiportfolio Organization Unit (Orgn Unit)

The use of an assessment instrument at multiportfolio organization level based on a comprehensive set of competencies with respect to project performance at organization level will act as an important step towards better belittling of the complexities and nuances of a professional organization.

The multiportfolio Orgn unit score for an organization establishes its current level of performance by identifying the gap in performance determinants. Though it is not intended to be prescriptive in terms of the gaps, but it highlights the areas with least scores showing room for improvement that can be aligned with the organization's maturity model for the success attainment.

3.6.1 Process flow for VDPI hardware interface

The process of integrating the VDPI tool with the hardware interface lies in collection and processing of information as a whole and in a much convenient manner with the ease of processing of information. The hardware device interface and their functions are represented in Table 10.

The initial stage of the device begins with the evaluation of any individual at project level, which requires the individual to input the field data of the project with reference to his/her roles and responsibilities as defined in the project charter. The interface is termed as IU. The personal competencies of individual are highlighted based on the observations gathered together using project information. But the IU interface is limited to self-assessment only and does not include individual benchmarking using VDPI tool.

Based on the determination of performance indicators and their determinants with respect to the IU, these units can be redefined. Further, an individual can score oneself on the basis of the work executed towards each project performance indicator of the project, which provides the IU with their own VDPI score as per the carried out self-assessment. This data of the carried-out assessment is stored in the main server of the app.

Table 10 Hardware device interface and their functions

Hardware device	Hardware device function
IU	Individual self-assessment, Individual bench marking
PH	CPM assessment of project,
	CPM benchmarking on project, Remodelling of VDPI on project
OU	CPM assessment of project,
	CPM benchmarking on project, Remodelling of VDPI on project
Orgn Unit	Organizational level assessment of CPM,
	Multiportfolio organizational-level remodelling

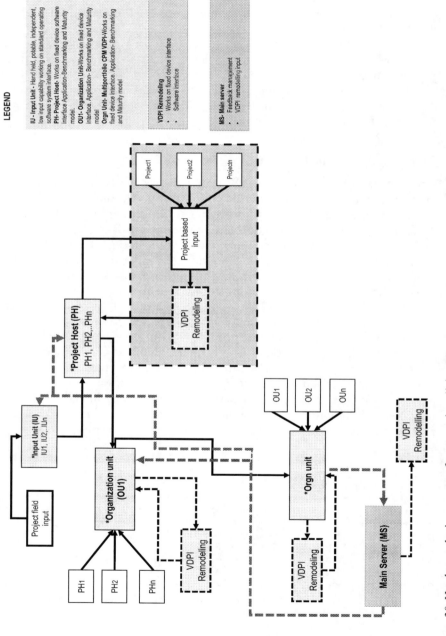

Figure 20 Hardware device support for operationalizing VDPI.

Performance index for construction project managers   77

This allows the individual to assess one's performance without any bias. The obtained scores are further identified as below expectations, meets expectations, and above expectations. Suitable development plans can be adopted to improve their performance. Further, the individual scores can be re-evaluated based on the remodelling approach adopted and can be re-accessed after implementation of the development plan. At the next level, it is the project host who evaluates his/her performance using the VDPI tool for their respective projects (being handled by them) and the input data is majorly related to the project-based field input. The end result for this stage is the use of self-assessment and maturity model through the application of benchmarking which ultimately reflects the CPM's self-assessment on project and CPM benchmarking on project.

The next stage is related to the organization level which is represented as OU units in the flowchart. This represents independent verticals within the organizations whose performance is assessed from the outputs as received from project hosts analysis as an input variable and the final results are organizational-level self-assessment and organizational-level benchmarking.

The last level of the VDPI hardware device technical interface is multi-portfolio organization level, and it requires the input from the multi-orgn units. The organization's performance level is evaluated based on the industry standards and other competing organizations. At multi orgn level, the major tools to be used are expert judgements, industry trend analysis, etc., which help an organization in assessing its level of performance with respect to others in the competition market.

Through the VDPI score, the organizational scores are obtained which helps the organizations in preparation of development plans related to enhancement of organizational performance by identifying the key performance indicators (KPIs) governing organizational performance and meeting critical organizational milestones.

The detailed flowchart representing technical support device hardware interface and data processing is represented in Figure 20.

3.7 Inferences

The VDPI tool encourages continual development as its methodology based on continuous and iterative evaluation process for a CPM both at individual and organizational levels in which the actual requirements are reviewed, competence is accessed, and based on the score, competency development is formulated. Each iteration of the competency measurement score is treated as individual's as well as organization's self-assessment and development programme which are in line with the organizational maturity levels. This helps in eliminating competency gap of CPMs. The gaps can be viewed holistically to give a multidimensional picture, or can be viewed individually to address specific development opportunities.

78 *Performance index for construction project managers*

The VDPI will act as an indicator for assessing the performance of a CPM. The derived variables for the equation for the quantification process are specific to the construction industry and are derived using the questionnaire survey approach. It identifies the key competencies of a project manager at different levels of hierarchy, which impacts the key determinants of a project's performance specific to the construction industry.

It even helps an organization to evaluate the performance of its individual units, combined together as a whole, which aids in better planning, management of an individual professional and organizational development collectively, for its application different levels of input units at individual level, project host at project level, orgn unit at organizational level, and for dealing with project programme level at multiportfolio organization level.

References

Alin, A. (2010). Multicollinearity. *Wiley Interdisciplinary Reviews: Computational Statistics, 2*(3), 370–374. https://doi.org/10.1002/WICS.84

Blalock H.M, J. (1963). Correlated independent variables: The problem of multicollinearity. *Social Forces, 42*(2), 233–237. https://doi.org/10.1093/SF/42.2.233

Grewal, R., Cote, J. A., & Baumgartner, H. (2004). Multicollinearity and measurement error in structural equation models: Implications for theory testing. *Marketing Science, 23*(4), 519–529.

Gujarati, D. N., & Porter, D. C. (2003). *Basic econometrics* (4th ed.). New York: McGraw-Hill.

Jensen, D. R., & Ramirez, D. E. (2012). Variance inflation in regression. *Advances in Decision Sciences, 2013*(1), 1–15.

Lavy, S., Garcia, J. A., & Dixit, M. K. (2014). KPIs for facility's performance assessment, Part II: Identification of variables and deriving expressions for core indicators. *Facilities, 32*(5), 275–294. https://doi.org/10.1108/F-09-2012-0067

Malhotra, N. K., Peterson, M., & Kleiser, S. B. (1999). Marketing research: A state-of-the-art review and directions for the twenty-first century. *Journal of the Academy of Marketing Science, 27*(2). https://doi.org/10.1177/0092070399272004

Maruyama, G. (2014). Basics of structural equation modeling. In *Basics of structural equation modeling.* https://doi.org/10.4135/9781483345109

Project Management Institute. (2017). *A guide to project management body of knowledge (PMBOK guide): Vol. Sixth edition.*

Siegel, A. F., & Wagner, M. R. (2022). Multiple regression. *Practical Business Statistics,* 371–431. https://doi.org/10.1016/B978-0-12-820025-4.00012-9

Unegbu, H. C. O., Yawas, D. S., & Dan-asabe, B. (2022). An investigation of the relationship between project performance measures and project management practices of construction projects for the construction industry in Nigeria. *Journal of King Saud University - Engineering Sciences, 34*(4), 240–249. https://doi.org/10.1016/J.JKSUES.2020.10.001

Verbeke, W., & Bagozzi, R. P. (2000). Sales call anxiety: Exploring what it means when fear rules a sales encounter. *Journal of Marketing, 64*(3). https://doi.org/10.1509/jmkg.64.3.88.18032

4 Value-driven performance assessment of construction project managers

4.1 Introduction

The previous chapter discussed about the concept of VDPI for evaluating the performance of CPMs. These drivers are the performance indicators and their determinants which have been identified in the current chapter of this book. The sections covered in the chapter include the need for the identification of performance indicators, traditional project performance indicators, use of performance indicators, the performance indicators specific to construction industry for defining the performance of CPM, and processes of project management based on which the determinants of project performance are derived with the constraint of the success of a construction project. These determinants have been derived using the existing literature and industry practice standards related to project management.

4.2 Performance indicators

4.2.1 Need for performance indicators

Indeed, there appears to have been a lack of a formal project appraisal process for construction projects. There doesn't seem to be a consensus on how to define the success of any construction project or its performance indicators, making them a vaguely defined term.

In the case of construction industry, the success of a project is governed by various factors as it involves a number of parties, different phases, stages of work, etc. (Takim & Akintoye, 2002).

To successfully complete any project, it is imperative to identify performance parameters. *Performance indicators are a means for defining the quantifiable data required to demonstrate that a planned effort has produced the desired result or intended outcome.*

They are the measurable evidence for proving that the efforts have been taken to achieve the desired results. The performance indicators are emerging as industrial jargon and are being used as performance measurement tools.

DOI: 10.1201/9781003322771-4

80 *Value-driven performance assessment*

Thus, performance indicators are a means to quantify the planned strategy which has been executed and levelled up to which the desired results have been achieved. These measures are tried to be made precise, leaving behind any kind of ambiguities in measure indicators.

The objective behind performance indicators is to measure and identify opportunities for improvement. It ultimately helps in the overall performance evaluation of an individual/project/organization and in the formulation and quantification of the strategic performance of an organization (Yang et al., 2010). Performance indicators in hierarchal order are generally formulated into assessment frameworks. To define success measures, it is important to identify the key success parameters which are related to both individual as well as organizational excellence.

The identification of performance indicators is extremely important as it depends on the knowledge of organizational goals and project's success parameters. It is also associated with the techniques that are being implemented and used to examine the status of any construction project and its critical activities, which have a significant impact on the overall project progress and thereby motivate an individual/organization to improve. The definition and parameters of success will be different with respect to every individual, project, organization, and industry, so the performance indicators should be wholehearted with respect to overall success goals. The organizations will also benefit through this and help in better decision-making as the measured performance indicators would provide a better picture of the organizational progress towards achieving its goals. Different performance assessment frameworks have been created in the management literature as a reaction to the requirement for ongoing improvement (Takim & Akintoye, 2002).

The traditional performance measures are not sufficient enough to gauge a project manager's performance, as they lack in providing the appropriate information that they need towards stimulating their continuous professional development (Ahadzie et al., 2008a). The unique nature of construction projects also governs these performance measures and their level of relevance in gauging the CPM's performance. The CPM's actual performance gap is the corollary of the evaluation carried out using these indicators. It has been widely accepted that one of the main goals of performance indicators is to serve as standards for encouraging the development of best practices (Barber, 2004).

4.2.2 *Meaning of performance indicators*

Performance indicators are the indicators intended for measuring progress towards any desired results. They help in forming the analytical basis and are very crucial for determining what needs to be measured to judge the performance.

The following features govern the choice of performance indicators:

Value-driven performance assessment 81

- Observable proof of movement in the direction of a goal
- Measure what needs to be measured to make better decisions
- Provide a comparison that evaluates how much performance has changed over time
- Has the ability to monitor project performance, people performance, economics, effectiveness, timeliness, budget, compliance, and behaviours (*What Is a Key Performance Indicator (KPI)?*, n.d.).

A performance indicator can be identified considering the SMART criteria (Vyas Gayatri & Kulkarni Saurabh, 2013):

- S – Individual/Project/Organization specific
- M – Measurable
- A – Achievable
- R – Relevant to the success of individual/project/organization
- T – Time phased (outcomes should be shown for a relevant period)

The inputs for defining the performance indicators of any CPM will be based on the performance parameters of the industry requirements. The performance indicators can be qualitative (qualitative indicators define the characteristics of a business decision and process) as well as quantitative (qualitative indicators can be continuous as well as discrete).

The performance indicators are divided into two categories. These indicators being the leading and lagging indicators are represented in Table 11.

- Leading indicators – these indicators predict the outcome of any process and confirm its long-term trends. These types of indicators are

Table 11 Concept of use of leading and lagging indicators

Leading indicators		Lagging indicators		Remarks
Targets	*List of indicators*	*Targets*	*List of indicators*	
Newly implemented project management process	Time, cost, scope, quality, etc.	Traditional project management processes	Time, cost, scope, quality, etc.	The sub-indicators for the identified indicators might vary for each target.
New technology	Time, cost, scope, quality, etc.	Old technology usage	Time, cost, scope, quality, etc.	

82 *Value-driven performance assessment*

suitable for predicting the after-effects of a launch of a new product in the market/strategy in the organization.

- Lagging indicators – these indicators are used to measure the outcome of an action undertaken to reflect the achievement or fiasco of action and help in analyzing the key impacts of necessary actions (Kagioglou et al., 2001).

Hence, in this chapter, we will discuss in detail the various indicators governing the performance of any CPM and the broader categories in which they have been grouped are time, cost, design, scope, and contract.

4.2.3 Traditional project performance indicators

The traditional method of measuring any project's performance is indicated by the completion of any project as per the planned budget, timeline, and meeting client expectations. As construction projects are getting more complex, there's a growing need to identify the key project performance indicators in a very standardized manner to promote their wider application. The traditional indicators of time, cost, and quality cannot only be the principal factors intended for analyzing any project's performance (Chan & Chan, 2004b).

The crucial activity for assessing any project's performance is to evaluate the performance of each participant involved in all phases of a project, and to prioritize them by determining the extent of any project's success and achieved targeted improvements. The idea behind the identification of performance indicators is to link together success with routine tasks at individual as well as organizational levels.

In order to exit and emulate the dynamic market conditions, it is essential to continuously enhance and update the skill set with the emerging nature of projects and their needs. To identify and measure the growth or level of improvement, it is essential to determine the performance indicators which are the key indicators for measuring performance. Performance measurement is something that lies at the heart of ceaseless improvement and is associated to offer the facility with the development of direction, traction, and speed of an organization (Luu et al., 2008b).

The establishment of a measurement approach for too many performance indicators is very tough and requires the identification of key performance indicators concerning to the core objectives for attaining any project's success related to project management. As a traditional approach, project success is unified with project performance that defines the iron triangle of project performance, which states that project success is based on the scope, time, and cost parameters of a project.

Project success is an abstract concept and it is complex to define the criteria for a project's success (Chan et al., 2002) The traditional indicators have their limitations; some academicians and researchers have

Value-driven performance assessment   83

pointed out that the iron triangle alone cannot suffice as the prerequisite for assessment of any project's success and this idea of lack of project performance attributes in the iron triangle of project management is embraced by the global project management community. But the parameters of the iron triangle should always be included with newly formulated definitions of project performance. As construction projects are subjected to the dynamic, evolving environment, the complexity of construction projects leads to their uncertain nature, which is a contributing factor to emerging behaviour of construction projects. Hence, the identification of performance indicators for defining the success of the projects is a challenging task (Orihuela et al., 2017).

The identification of performance indicators for construction projects can be formulated using the following research questions:

- What is project success according to construction projects?
- Which are the crucial indicators governing any construction project's success?
- Which performance indicators can be used to assess how well construction projects are performing?

Key performance indicators are the leading indicators that are derived from the fundamental characteristics of any project and are listed in Table 12. They govern the success of a project and affect the overall project outcome (*CII- RT008*, 1987) (Takim & Akintoye, 2002) in their research work identified. Majorly, *construction time, cost, predictability of costs, defects, safety, customer satisfaction with the product, and customer satisfaction with the service* are the seven project performance indicators. *Safety, profitability, and productivity* are the three company performance indicators.

4.2.4 Use of performance indicators

- Linking of goals to daily routine tasks
- A measure of checking the current status of performance/progress
- Helpful in the formulation of growth plans
- Aid decision making
- Meaningful data collection
- Tracking of progress to ensure that the project is moving in the exact direction.

The Project Management Institute defines project as *"discrete but multidimensional activities that serve as vehicles of change"*. Project success is defined by project completion with respect to timely completion, within the approved budget, as per the specifications, and accomplishment of business goals with respect to project, which might be helpful in defining project success related performance indicators (Bannerman, 2008).

84 *Value-driven performance assessment*

Table 12 KPIs considered in past studies for evaluating the performance of construction projects

S. No.	Key Performance Indicators (KPIs)	References
1	Construction Cost Performance, Time of construction, Customer satisfaction on services, Customer satisfaction on products, Quality Management System (QMS), Project team performance, Change Management, Material Management, Labour Safety Management	(Luu et al., 2008a)
2	Deviation from cost, construction due date, change in scope of awarded work, safety- accident rate, risk rate, efficiency of direct labour, construction- productivity performance,	
3	Results: Cost- cost variation, Time- Schedule variation, Quality- cost of client claims (cost of repairing claims/ defects, number of claims), Scope- Change in contract scale, Safety – accident index, risk rate, Labour- efficiency of direct labour	(Alarcón et al., 2001)
	Process:	
	Construction-productivity output (monthly sales/ monthly man-hours sold), rework, waste, transportation, Procurement- Urgent orders (number of urgent orders/ total number of orders), cycle time, mean delay time, Planning- Effective planning (%planned complete), Organization management- administrative productivity (cost of general administration/ monthly sales), Design: quality of design, design errors.	
	Variables:	
	Work force-Training, Sub contractors-subcontractor ratio (subcontracted cost/total cost)	
4	• Deviation of cost by project, • Deviation of construction due date, • Change in amount contracted, • Accident rate, • Risk rate, • Efficiency of direct labour, • Productivity performance, • Rate of subcontracting, • Client cost complaints, • Urgent orders, • Planning effectiveness.	(Markovic et al., 2011)
5	• Construction cost • Construction time • Cost predictability • Time predictability • Defects • Client satisfaction with the product and services • Safety • Profitability • Productivity	(Takim & Akintoye, 2002)

Value-driven performance assessment 85

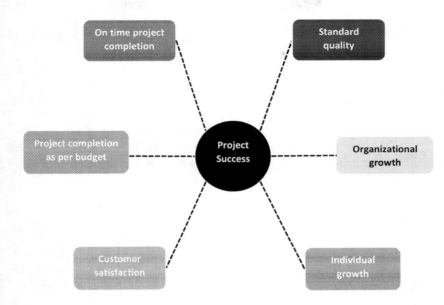

Figure 21 Dimensions of project success.

The dimensions of project success for defining success criteria are represented in Figure 21.

Different processes and decisions implemented during the entire lifecycle of a construction project derive the final results of a project. Granted, CPM's competency and project success are interlinked and well established (Ahadzie et al., 2008a). Certain studies suggest dividing and categorizing performance indicators as enterprise performance, project performance, and benchmarking programs (result, process, and leading indicators) (Alarcón et al., 2001; Orihuela et al., 2017).

The indicators can be defined into three categories:-

- Results – Indicators that are used to measure final results related to any projects success, such as project time, cost, scope, and quality.
- Processes – Indicators that are used to measure the performance of processes embedded during the life cycle of a project, such as procurement, design, planning, etc.
- Decisions – Variables and strategies that are not directly related to the processes involved but have an impact on the project performance. For example, type of project, type of contract, etc.

The performance indicators are to be selected in such a way that they are relevant to the topic being addressed and are extremely precise so

86 Value-driven performance assessment

as to allow for a detailed understanding of the indicator with project performance (Meade, 1998). The selection of performance indicators and the final analysis process for helping in management decisions is represented in Figure 22.

Construction industry is a location-specific industry and no two projects can be same, which hinders the project performance (Garnett & Pickrell, 2000). The radical verification of consideration for key project performance indicators with respect to construction industry is needed to be detailed out in a very subjective manner. Further, as identified, these indicators are generally related to project's time, cost, scope, design, and contract, which have been identified by considering both the concepts of the project lifecycle and project phases (Orihuela et al., 2017).

> In the next section of this chapter, a radically finalized list of project management performance indicators has been listed down by mapping them with the end user goal satisfaction, methodology, and practices adopted in the construction industry. Further, it is detailed with respect to the key determinants defining a project's success indicators to maximize efficiency and minimize the associated threats as represented in Figure 23.

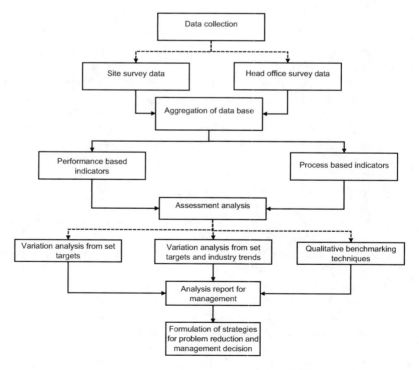

Figure 22 How performance indicators support management actions.

Value-driven performance assessment 87

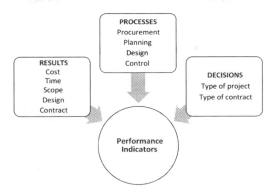

Figure 23 Performance indicators of construction projects.

a   Time
b   Cost
c   Scope
d   Design
e   Contract

The significance of identified performance indicators and their determinants for construction project's success might vary with the advancement in technology and development processes.

These performance indicators are generally analyzed with respect to three broad terms in the present body of project management knowledge. In the previous chapters, they have been represented as skill and competence-based performance, that can be both tangible and non-tangible, considering the dynamic nature of onus needed from any CPM are related to the following:

- Job focused Competence
- Personal focused
- Role focused

4.2.5 Time management performance indicators and determinants

The term "time" in any project is referred to as a synonym for project schedule, timeline, baseline, and deadlines in the construction industry. "Time" is a broad term encompassing scheduling time, lead time, delivery time, extension time, etc. The factor of time is crucial and its increasing salience is predominant in the existing project management body of knowledge. Also, it defines the success of any project by completion of project within the stipulated time, "time" being the essence of completion of any project (Claessens et al., 2007).

88   *Value-driven performance assessment*

As per IS code 15883 Part (2), time management is crucial for construction projects as they are generally aimed to be completed in the allocated time. The allocated time frame for any project is what matches the time frame of the user/client based on the requirement. In many of the cases, the feasibility of a project is governed by the time of completion of the project. Therefore, it becomes essential to develop, optimize, and manage a project's schedule on a realistic basis considering every aspect of a project. Project time management cannot be seen in isolation as a single aspect and requires a CPM to understand the interdependence of various construction management processes. It is the responsibility of CPM to make sure that the project is completed within the planned duration of the project as agreed upon by the stakeholders. As a CPM, time management needs to viewed in a holistic sense comprising other dimensions of construction project management performance like time, cost, quality, scope, design, etc. (BIS, 2013).

According to previous studies conducted by various researchers and available project management standards, there are different terminologies and formulae for defining and analyzing the term "time" used in the field of project management such as construction time, speed of construction, time variation, schedule performance index (SPI), and schedule variance (SV) (Chan & Chan, 2004a). Construction time is the absolute time that is calculated as the number of days/weeks from the start on site to the practical completion of the project. Speed of construction is the relative time, which is defined by gross floor area divided by the construction time. Time variation is measured by the percentage of increase or decrease in the estimated project in days/weeks, discounting the effect of the extension of time (EOT) granted by the client (Reichel, 2006). Another way to look at time performance is through SPI. Based on the theory of earned value management, SPI is a measure of the schedule efficiency of the project; SPI is determined by dividing the earned value by the scheduled value (Nassar, 2009) (Project Management Institute., 2005). Any value of SPI < 1 indicates that the project is running behind the schedule, SPI =1 indicates that the project is running as per the schedule, SPI>1 indicates that the project is running ahead of schedule. Similarly, the term schedule variance is used in analyzing time performance. It indicates the value of work that is ahead or behind the planned schedule which is defined as the difference between earned value and planned value (Project Management Institute, 2005). Discounted construction time is defined as the difference between the actual construction time and revised construction duration is known as discounted construction time (Luu et al., 2008b).

A consolidated list of aggregated tools and techniques based on the knowledge area of project schedule management and its processes (initiating, planning, executing, monitoring and control and close out) as described in PMBoK has been represented in Table 1 which are needed to be encapsulated in a project schedule management plan.

Value-driven performance assessment 89

Table 13 Tools and techniques of project performance related to time as per PMBOK@ Guide

Project management process	Tools and techniques
Time Performance	Decomposition
	Expert time judgement
	Rolling wave planning
	Analogous estimating
	Parametric estimating
	Three-point estimating
	Reserve analysis
	Critical path method
	Critical chain method
	Resource levelling
	What-if scenario analysis
	Schedule compression

Table 14 Mapping of time management processes and determinants

Time performance management process and determinants

Sources	Process groups	Processes	Determinants
(Ahadzie et al., 2008a)	Planning process group	Plan Schedule Management	Planning work coordination
(Nassar, 2009)		Define Activities, Develop schedule	Coordinated schedule development
(Oliveros & Vaz-Serra, 2018)		Sequence activities	Risk forecasting
(Pariafsai & Behzadan, 2021)		Estimate activity resources	Effective resource planning
(Yang et al., 2010)		Estimate activity durations	Construction time, Speed of construction
(Chen et al., 2008) (Luu et al., 2008)	Monitoring and controlling process group	Control schedule	Effective schedule control, Time variation

4.2.5.1 Determinants of time management performance

Though there are various methods, tools, and techniques available for analyzing the time performance of any project as defined in the PMBoK Guide (Project Management Institute, 2017a) yet none of the available standards provides one with a holistic approach to determine the time management performance of a CPM with respect to project success in case of the construction industry. Based on the review of existing literature, a few of the key determinants for assessing the time management performance of a CPM have been identified and are listed in the next section of this chapter and in Table 14.

90 *Value-driven performance assessment*

The major time performance determinants which impact the overall project performance and require better monitoring and control by the CPM are listed in Table 15 along with their brief description. The identified determinants of time management performance have different weightages under varying scenarios like typologies of project, project objectives, etc. which might vary or require few additions or subtractions based on the project's requirements.

i **Planning work coordination** – It is related to the development of a coordinated project schedule involving a holistic sense of each activity required in bringing out the final product for work packages considering the scope of work to be carried out as per the agreed contract. Coordinated schedule development also involves the preparation of a schedule with coherence to project stakeholders involved. The stakeholders generally include all members of the project team, sponsors, consultants, contractors, clients, stakeholders coming from governance, etc. As stakeholders generally influence the overall project deliverables as a CPM, it is essential to pay attention to stakeholder requirements and obtain their coherence in developing a coordinated project schedule for achieving better efficiency. Development of the schedule should involve analysis of the sequence of activities, durations, resource allocation, and schedule constraints with respect to all stakeholders. Especially, in case of construction projects and their dynamic nature, proper linked information flow within the stakeholders is established based on which coordinated schedule is developed.

ii **Effective schedule control** – Controlling of schedule is related to the monitoring and control phase of the project lifecycle. It ensures proper monitoring of project activities status, which involves updating of the project schedule as the project progresses. It is the responsibility of a CPM to make sure that the project progress is tracked and effectively monitored with respect to the set baseline while ensuring the effectiveness of schedule control with respect to the project deadline and removing hypothetical situations related to schedule control. The project manager is responsible for ensuring adherence to schedule as agreed in coherence with the project stakeholders. Further, it is ensured that the schedule is being followed during each stage of the project through proactive and reactive measures as required (BIS, 2013). It is the duty of the CPM to carry out delay analysis which focuses on analysis at each activity level to determine where, when, and why the delay occurred and identification of the responsible stakeholder, and if the forecasted delays can be controlled using strategic planning. The cost of delay calculation and its communication with the respective stakeholders has to be carried out by the project manager. The use of adequate project schedule monitoring and control techniques by the project manager might help in controlling project delays.

iii **Risk forecasting** – Based on the underlying scheduling data available, it is the responsibility of the project manager to identify and analyze the potential risks associated with the project's critical path or at any other stage of the project which are quite common in construction projects, considering the uncertainty present in the scheduling data. The forecasting of risks can be carried out using both quantitative and qualitative processes. The forecasting of scheduling risks would help in better management of the identified risks by incorporating the estimated risks and their impacts in the schedule itself. Any misfit scenario should be analyzed and reported by the CPM to the relevant stakeholders for managing the forecasted risk.

iv **Effective resource planning** – The CPM is responsible for ensuring that an adequate number of resources (manpower, material, plants, and equipment) are allocated, known as resource scheduling, which is based on the timeline set for each activity completion. The CPM must ensure that no additional resources are being allocated for an individual activity which might be just about adding an extra cost to the finalized project budget by focusing on resource optimization techniques. The project manager is responsible for managing the resource tracking to avoid variation in terms of planned resources as per the project baseline plan. Regular monitoring of deployed resources is also essential to determine the variance, or any additional resource mobilization as and when required. The success of a construction project significantly depends on the resource.

Table 15 represents the time performance determinants as they aggregate in proportion of their weight towards overall time performance. A pictorial representation of time management performance determinants and their applicability with consideration to PMBoK process groups is represented in Figure 24.

4.2.6 Cost management performance indicators and determinants

The cost of a project is an important indicator of any project's success. In case of construction projects, a broad range of stakeholders are involved, and the major focus of stakeholders is on the cost of the project. The literature

Table 15 Determinants of time performance

Time performance	W1	Weightage	Determinants
		W11	Planning work coordination
		W12	Effective schedule control
		W13	Risk forecasting
		W14	Effective resource planning

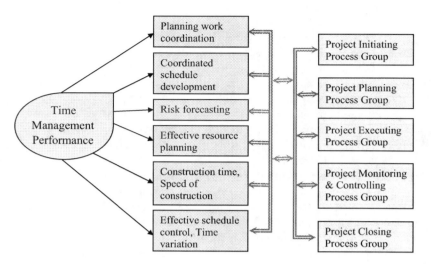

Figure 24 Determinants of time management performance and PMBOK process group.

suggests that there are various ways through which cost performance can be measured like cost per unit, total cost of construction, variation cost, etc. (Chan et al., 2002). The main objective of cost management is to ensure project completion within the authorized budget as agreed by the concerned stakeholders. Cost management involves refining the cost, cost budgeting/ estimation, cost management, resource planning, and cost monitoring and control (BIS, 2009).

Project success in terms of cost can be defined as the completion of project within the budgeted cost. There are multiple indicators through which cost performance of any construction project can be measured, some of which are discussed in the subsequent section for measurement of cost performance. Cost of construction, being one of the indicators, is defined as a total actual cost of construction of the project. Similarly, unit cost is used in defining the cost performance of a project – the measure of unit cost is referred to define the cost per unit for the ease of comparison (Chan & Chan, 2004a). It is one of the basic methods of quantitative reporting of cost. Based on the unitary method of estimating, it offers a simplistic approach to productivity measurement (Vyas Gayatri & Kulkarni Saurabh, 2013).

Other terminologies used in project management standards related to cost management include cost of change (it is equivalent to the cost of works related to change orders) and net variation cost. The net variation cost is defined as the difference in total budgeted approved cost and the actual cost after completion of project (Chan & Chan, 2004a). This terminology is based on the concept of job cost reporting as described by Vyas Gayatri and Kulkarni Saurabh (2013).

Based on the concept of earned value management, cost performance index (CPI) is defined for assessing the cost performance of a project. Cost performance index is a measure of the cost efficiency of a project. As per Project Management Institute (2005), any CPI value >1 indicates favourable condition, whereas CPI value <1 indicates unfavourable condition. Another way to look at cost performance is the cost variance, which highlights whether the project is on budget or above budget. It is determined as the difference between actual cost and earned value. A CV value >0 indicates that the project is under budget, cost variance (CV)=0 indicates that the project is as per the budget, and CV<0 indicates that the project is over budget.

Another index used for assessing the cost management performance of a project is Billing Performance Index (BPI). It is a measure for determining the efficiency of invoicing the client for the earned work. The BPI is determined by dividing the Billed Revenue by the Earned Revenue for the Work Performed (Nassar, 2009). It helps in better management of project cashflow. In similar terms, Profitability Performance Index (PPI) is used as a measure to determine how profitable the project is to date. The PPI is determined by dividing the Earned Revenue of the Work Performed (ERWP) by the Actual Cost of the Work Performed (ACWP) (Nassar, 2009).

The *KPI Report for The Minister for Construction* (1999) highlights other cost-related indicators to be considered such as cost-in use (annual operation and maintenance cost) and cost of rectifying defects during the maintenance period also.

A consolidated list of aggregated tools and techniques based on the knowledge area of project cost management and its processes (initiating, planning, executing, monitoring and control, and close out) as described in PMBoK has been represented in Table 16 which are needed to be encapsulated in a project schedule management plan.

Table 16 Indicators of project performance related to cost as per PMBOK@ Guide

Project management process	Tools and techniques
Cost Analysis	Expert cost judgement
	Analogous estimating
	Parametric estimating
	Three-point estimating
	Bottom-up estimating
	Reserve analysis
	Earned value management
	Forecasting
	To-complete performance index

94 Value-driven performance assessment

4.2.6.1 Determinants of cost management performance

The major cost performance determinants which impact the overall project's performance and require better monitoring and control by the project manager are described below. The identified determinants of cost management performance have different weightages under varying scenarios like typologies of project, project objectives, etc. which might vary or might require few additions or subtractions based on the project requirements.

Based on the review of existing literature, few of the key determinants for assessing cost management performance of CPM have been identified and mapped against the key are listed in Table 17. These determinants are mapped against the knowledge areas and processes of management as mentioned in PMBoK guide, to establish cost management perspective and process area during the project lifecycle.

i   **Effective cashflow management** – It is the responsibility of the project manager to ensure uninterrupted cashflow as per the authorized project cashflow baseline with respect to each milestone as defined in the project schedule. The project manager should ensure that well-coordinated and updated cashflows are developed with respect to the liabilities of individual stakeholders. A CPM must adapt in the planning of cashflow of the project so that the cashflow requirements for each stakeholder are met and it does not lead to disruption of work planned as per the schedule.

ii **Controlling budget variance** – The CPM should monitor the project cost as per the authorized budget to avoid any kind of budget variance

Table 17 Mapping of cost management processes and determinants

Cost management performance processes and determinants

Source	Process groups	Processes	Determinants
(H. A. E. M. Ali et al., 2013)	Planning process group	Plan cost management	Effective cash flow management
(Toor & Ogunlana, 2010)			Construction cost, unit cost
(Nassar, 2009)		Estimate cost	Controlling budget variance
(BIS, 2009; Paul & Basu, 2021)		Determine budget	Managing risk contingencies
(Nassar, 2009) (BIS, 2013; Skibniewski & Ghosh, 2009)	Monitoring and controlling process group	Control costs	Controlling cost overruns
			Change cost
			Cost performance index, Cost variance

Value-driven performance assessment 95

and ensure an iron grip on project cost. During the execution stage, different complexities arise which might lead to budget variance and require better management skills to avoid the actual cost variation from the authorized/planned budget for the assigned scope of work. The CPM must identify the factors responsible for variation in the authorized budget and take necessary actions to control any associated variances.

iii **Managing risk contingencies** – Contingencies are majorly associated with the project scope of work in which the estimates consider some extra loading with respect to a particular activity. As a CPM, one must ensure that the extra loading is within the limit of the project. As the project plan evolves, the project manager must account for the change, so contingency planning, monitoring, and management are necessary to avoid any disputes arising from the unknowns of the project. CPM must be well versed in mitigating the risks arising out of the unknown conditions which are not a part of the scope of work as per the agreed contract terms and conditions. Further, the contingency associated with each task/activity must be identified to ensure proper management.

iv **Controlling cost overruns** – The determinant of controlling cost overrun and budget variance is quite similar. The standard is to make sure that the project is completed within the approved estimated cost as per the agreed scope of work. The project's actual budgeted cost of work performed and planned cost of work performed need to be monitored by the CPM. The project manager controls cost overruns by bringing the expected cost overruns within the acceptable limits, if not eliminating cost overruns completely. Ideally, cost should justify utilization of contingencies in response to the prevenance of risks and therefore, must be less than the total projected cost. Thus, oversight of risks and their cost implications is an important consideration while strategizing initiatives controlling cost overrun.

Table 18 represents the cost performance determinants as they aggregate in proportion of their weight towards overall cost performance and their applicability with consideration to PMBoK process groups as represented in Figure 25.

4.2.7 Scope management performance indicators and determinants

Scope is one of the vertices of the iron triangle of project management along with time and cost. It represents one of the major constraints in relation to project success. It refers to *"all the work involved in creating the deliverables of the project and the processes used to create them"* (Jainendrakumar, 2015). The scope of work of any project is directly linked with the project objective and defines the goals of the organizations and final outcome to be achieved/ deliverables to be delivered, as it relates to the work content to be delivered.

Table 18 Determinants of cost performance

Cost performance	W2	Weightages	Determinants
		W21	Effective cash flow management
		W22	Controlling budget variance
		W23	Managing risk contingencies
		W24	Controlling cost overruns

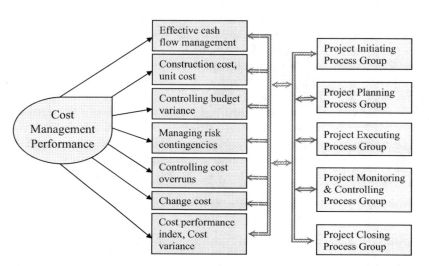

Figure 25 Determinants of cost management performance and PMBOK process groups.

It is one of the most pivotal factors for defining any project's success and needs to be integrated with the objectives (Mirza et al., 2013).

The concept of scope control has significance in terms of managing scope changes. The associated impacts of scope change need efficient and effective overall control of the project, otherwise they will have a cascading effect on time and/or cost performance. For the multiple stakeholders involved in any project, scope of the project is a key factor in determining its primary function. These different investors/stakeholders have their interest in short-term or long-term implications of project, especially during the operational stage. As rightly identified by various researchers, poor scope management is a strong reason for the overall failure of construction projects as changes in scope of project can have detrimental effect questioning the very feasibility of the project (Nahod, 2012).

Studies conducted by various researchers have pointed out that it is essential to document the project requirements before proceeding with the other phases of any project during its initial stage, which is called scope documentation. The process of scope documentation stands out to be necessary

Value-driven performance assessment 97

in defining the baseline of the project scope and getting it approved by the relevant stakeholders before setting the project. It is a part of scope documentation, and in future it provides a baseline for determining any kind of changes in scope, evaluation of actual work done, and any decisions to be taken (Cockfield, 1987). Effective stakeholder involvement is the key to defining and managing scope from the earliest project development stage. While the experts and design professionals undertake project typology-specific processes to identify scope, the stakeholder/investor/client empowerment to play a proactive role is warranted. In any case, it helps in due recognition of their specific concerns and elimination of opportunities for manifesting the same at an advanced stage of project evolution, leading to expensive reworks.

As per AIPM (2021), scope definition may include the following:

- Project Objectives
- Product Scope Description
- Project Requirements
- Project Boundaries
- Project Deliverables
- Product Acceptance Criteria
- Project Constraints and Assumptions
- Initial Project Organization
- Initial Defined Risks
- Schedule and Cost Factors
- Work Breakdown Structure

As per AIPM (2021), considering the scope management area of project management, the knowledge and skills required from a project manager would be:

- Planning
- Monitoring and tracking control
- Teamwork and communication skills
- Critical thinking, accuracy, attention to detail

Table 19 Tools and techniques of project performance related to scope as per PMBOK@ Guide

Project management process	Tools and techniques
Scope Management	Requirement
	Expert scope judgement
	Product analysis
	Alternatives identification
	Work breakdown structure
	Inspection

98 *Value-driven performance assessment*

PMBOK identifies various tools and techniques that may be used in scope management processes as represented in Table 19.

4.2.7.1 Determinants of scope management performance

The determinants from the point of view of scope management performance, in discharge of the functions of scope management are identified in the following section.

1 **Clarity of contract in scope/scope definition**
   In a study by Park (2009), clarity of contract was highlighted as a major performance attribute for any project. If project requirements are not mentioned adequately, it might lead to misinterpretation of the scope of work by the project team, which eventually leads to lack of accountability within the project team. Undefined goals and expectations later contribute to scope creep. A very important criterion in project control is always the scope definition.

2 **Change in project scope**
   The term scope creep is generally used to define the change in the scope of work that manifests to additional work. With the growing complex nature of construction projects, some uncertainties remain associated with the project, which turn out to be inevitable during the early stages of the project and lastly prime towards scope creep.
   Scope creep starts consuming the project's progress and finally dooms it to failure. It is the responsibility of a CPM to judge and communicate to other project stakeholders on whether the scope change request needs to be implemented or not.

3 **Omission versus incomplete scope of work**
   Lack of sufficient information in project contract documents and lack of proper details in the bidding package of the design details and specifications as required might lead to scope changes in further stages of the project lifecycle. This could lead to claims and disputes between the stakeholders. It is a fact that complete information may never be available for the CPM to define the scope accurately. However, the project manager is expected to identify such loose ends and minimize implications. Referred to as "rolling wave planning", the project manager must deal with incomplete information with foresight. This must not be confused with complete oversight of deliverables reflecting lack of project requirements in a given typology. The inability to identify a particular requirement and not having sufficient information about a particular requirement are two different situations, with the former being a failure in responsibility of the project manager.

4 **Scope planning**
   As per PMBoK, scope management has six processes out of which scope planning is the foremost that encompasses creating a scope management

Value-driven performance assessment 99

plan. Other project management guidelines, such as BIS also prescribe processes for strategizing planning for scope identification, monitoring and control, etc. through a documented scope management plan. Needless to say, this responsibility of CPM is dictated by the project peculiarities even though the overall structure or set of processes may be generic, hence the challenge for the CPM. Prima facie it may appear to be a team effort resulting from collective wisdom of all stakeholders and failure may be possible to be attributed to any scapegoat in the team. However, the project manager alone can ensure the success of robust scope planning that may not be vulnerable to scope creep subsequently. The performance determinants therefore must address this subtle distinction in the proactive role of CPM.

5 **Work breakdown structure (WBS)**
The work breakdown structure is an organized way to decompose the deliverables and identify the lowest unit of deliverable as a work package. Although it may appear to be an elementary exercise, it has nuances specific to the deliverables. There is no clear consensus of an ideal WBS for a given typology and peculiarities of project-specific responsibilities. Thus, the CPM reflects strategy to deliver project scope through WBS based on his/her wisdom. Elsewhere, WBS is also fundamental for the discharge of time-related responsibilities as well as commensurate resource allocation exercise. The CPM must employ WBS not only in scope management related but in respect of time and cost performance as well. Conversely, inaccuracy and inappropriateness of application of WBS would result in collateral performance failure in time, scope, and cost performance.

Based on the review of existing literature, some of the key determinants for assessing the scope management performance of a CPM have been identified and mapped against the key as listed in Table 20. These

Table 20 Mapping of scope management processes and determinants

Scope management performance processes and determinants

Source	*Process groups*	*Processes*	*Determinants*
(Park, 2009)	Planning process group	Plan scope management	Coordinating scope planning
(AIPM, 2021)		Collect requirements	Effective stakeholder involvement
(Mirza et al., 2013)		Define scope	Clarity of scope
(Nahod, 2012)		Create WBS	Strategic planning of work packages
(Mohsini & Davidson, 1992)	Monitoring and controlling process group	Validate scope	Sufficient information availability
(Loosemore & Muslmani, 1999)		Control scope	Controlling scope creep Monitoring project deliverables

100   *Value-driven performance assessment*

determinants are mapped against the knowledge areas and processes of management as mentioned in PMBoK guide, to establish scope management perspective and process area during the project lifecycle.

The determinants from the point of view of scope management performance, in discharge of functions of scope management are elaborated in the following section. The determinants of scope management performance have varying significance under different scenarios like typologies of project, project objectives, etc. which might vary or might require certain additions or subtractions based on the project requirements.

i **Coordinated scope planning** – The determinant of coordinated scope planning refers to the coordinated scope plan with respect to the project stakeholders as per the contract document, to ensure that each deliverable is added and timely delivered in accordance with project utilization plan consistent with the aspirations of stakeholders in a project. The CPM should ensure that the scope of work is known to the respective stakeholders and is being delivered, in addition to ensuring accomplishment of project objectives. Coordinated scope planning will eventually lead towards better understanding of how work needs to be managed during the execution stage with effective allocation of project resources. The scope of the project consists of the business planning process and the deliverables associated with the scope of the work. A well-developed coordinated scope plan becomes a major source of project success as it involves the integration of project time, cost, and resources with respect to project deliverables, with due consideration of project stakeholders.

ii **Effective stakeholder involvement** – The CPM must ensure that all the necessary stakeholders, both external and internal, are involved in the project based on their scope of work so as to avoid any conflicts which might arise in the future. The deliverables that are acceptable with respect to each stakeholder need to be fine-tuned which requires maintaining necessary interaction between the stakeholders. The CPM needs to focus on the establishment of the lifecycle management process involving stakeholder interaction at each stage. As a CPM, it is important to identify the stakeholders involved at different stages and their needs and expectations, as project success depends on stakeholders' satisfaction (Paul & Basu, 2021).

iii **Monitoring project deliverables** – It is the duty of the CPM to monitor all project deliverables with respect to individual stakeholders. It is essential to validate the project deliverables as reflected in the contract document specifications, etc. and establish clearly the acceptance criteria, thereby eliminating any discrepancies in communication as well as the documentation of the same. Monitoring of the deliverables relates to the correctness of the deliverables ensured through quality assurance and quality control compliances. The deliverables for each stage of the

Value-driven performance assessment 101

project should be defined and managed through proper monitoring and control mechanisms ensuring the fulfilment of the deliverables. The effectiveness of monitoring project deliverables is ought to be reflected through tangible and verifiable outcomes. The challenge for the performance of the project manager is regarding the reliability of efficacy of processes in the anticipation of desired results.

iv   **Controlling scope creep** – The uncontrolled expansion of project deliverables without adjustment to time, cost, and scope of the agreed project is termed as "scope creep" (Project Management Institute, 2017a). The phenomenon of scope creep can be externally and internally induced due to surrounding constraints. The CPM must ensure that the undertaken work is in accordance with the agreed business plan and/or project management plan in order to support effective change control and performance measurement processes and procedures. Regular monitoring of events that may lead to scope creep is preventive foresight that must be exercised by the CPM. Therefore, prevention of scope creep is a performance indicator while controlling the same is a virtue of lesser significance. Whenever a scope creep occurs, the project manager must try to protect the boundaries of the project's baseline; no requests must be accepted by the project manager until clear agreement between stakeholders for the scope creep (Abramovici, 2000). To avoid scope creep situations, pre-planning should be done with the incorporation of effective change management processes, despite the reality of rolling wave planning.

Changes in case of construction projects arise due to change in the agreed scope of work amongst the stakeholders, which might arise due to additional works or less works being executed based on the accepted scope of work. Each scope change has a direct implication on project's planned budget and schedule which ultimately leads to project disputes. Managing changes is linked with both monitoring of project deliverables and controlling scope creep which involves monitoring the status of project deliverables and management of changes to the project scope baseline. Since change is the only constant considering the uncertainties involved in construction projects it becomes essential to implement a change management process, focusing on avoiding scope creep and in case of necessary work ensuring proper acceptance by the stakeholders. The project manager must ensure the documentation of the agreed scope changes by the project stakeholders. While analyzing scope changes the CPM is expected to also evaluate reasons for not being able to anticipate first changes in the first place. Any oversight on the comprehensiveness of scope identification exercise must entail re-evaluation of scope identification which otherwise might manifest at another stage as a result of the similar circumstances that led to the trigger of the scope change itself.

Table 21 represents the scope performance determinants as they aggregate in proportion of their weight towards overall scope performance and their

102 Value-driven performance assessment

Table 21 Determinants of scope performance

Scope performance	W3 Weightages	Determinants
	W31	Coordinating scope planning
	W32	Effective stakeholder involvement
	W33	Monitoring project deliverables
	W34	Controlling scope creep

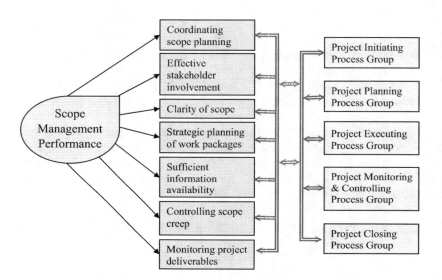

Figure 26 Determinants of scope management performance and PMBOK process groups.

applicability with consideration to PMBoK process groups is represented in Figure 26.

4.2.8 Contract management performance indicators and determinants

The dictionary meaning of contract is "*A written or spoken agreement, especially one concerning employment, sales, or tenancy, that is intended to be enforceable by law*".

The contract document is the central element of any project as it acts as the law book governing any project's execution procedure. Contract management is a broader term acting as a backbone of any project, which helps in running the business.

Contracts are notoriously difficult to evaluate in objective terms and are subject to different interpretations. Considering complicated language being associated with the contract as being a perfect contract is a hypothetical conjecture. Therefore, the role of contract management is integral to perform intellectual responsibilities of the CPM, thereby implying mutual coordination yet complying with binding framework. Contract management performance bears upon proactive decision making that can have adverse legal implications jeopardizing validity of the project itself. In such uncalled-for circumstances adjudication falls within the jurisdiction of arbitration and court of law where the CPM would not have any role to prevent outcomes. Therefore, all efforts at the hands of the CPM must be exercised as a part of contract management performance responsibility.

Contract management is a term encompassing the business case and establishes the procedure of management of relationships and review procedure for performance assessment (Guide *to Contract Management | CIPS*, n.d).

One of the major issues with contract formation is the formulation of contractual provisions. Explicating the role of framing in contract design is important as it helps us to understand why certain relationship outcomes are achieved. Contract management is an area of challenging concern for project management. Managing of the contract is more than simply writing the contract and following it during the project lifecycle (Harhad Meriem, 2018). Contracts are said to be of strategic importance as they can be a part of the solution to complex projects, which justifies the use of contract management in relation to risk management. Drafting of contract requires a deep understanding of the possible risks associated with the project role and responsibilities. Contract management is a granular process, having events occurring at each stage of project as represented in Figure 27.

Contract life cycle management is *"the process of systematically and efficiently managing contract creation, execution and analysis for maximising operational and financial performance and minimising risk"* (CIPS, 2007).

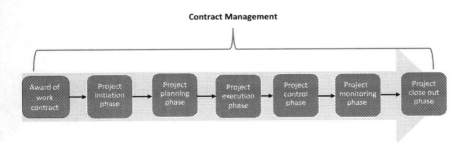

Figure 27 Contract management within a construction project lifecycle.

104   *Value-driven performance assessment*

4.2.8.1 Determinants of contract management performance

The performance attributes of any contract are directly linked with the success of the project. The key performance indicators of contract management would help project managers in better management of contracts by creating a checklist of performance indicators to be used for enhancing the project contract's effectiveness. The quantification of attributes of contract management might be more helpful in defining the contract conditions in a better way for the ease of contractual life assessment.

A contract document clarifies the conditions based on the project requirements and needs to be considered for defining project conditions, requirements, legal obligations, process of work execution, etc. in detail. The strategy of contract management should be coherent with the organization's procurement strategy. The contract's strategy should be developed with consideration of the following points:

- Business aims
- Critical success factors
- Time scaling and phasing
- Delivery capability

The determinants of contract management performance have been mapped against the project management process groups of procurement as defined in PMBoK guide and are represented in Table 22.

Table 22 Mapping of contract management processes and determinants

Contract management performance processes and determinants

Source	*Process groups*	*Processes*	*Determinants*
(Ahadzie et al., 2008b; BIS, 2009)	Planning process group	Plan procurement management	Risk sensitive procurement planning, independent estimates
(Chou & Yang, 2012; Unegbu et al., 2022)	Executing process group	Conduct requirements	Planning contractual obligations, negotiations
(Cho et al., 2009; Harhad Meriem, 2018)	Monitoring and controlling process group	Control procurements	Managing contractual obligations
(Project Management Institute, 2017a).	Closing process group	Close procurements	Claim administration Effective claim management
(BIS, 2009; Walker & Rowlinson, 2008)			Planning contract closeout

Value-driven performance assessment   105

Contract performance determinants can be related to the following time factors along with the cost associated with these factors:

- drafting and negotiation time
- number of versions and iterations
- contract administration time
- dispute settlement time
- contract cycle time

The determinants of contract performance with respect to the quality can be defined as:

- Degree of conformity to standards
- Degree of conformity to organizational goals, project objectives

Other determinants related to contract performance with respect to risk management can be defined as:

- Amount of agreement's expiry date
- Number of improper signature approvals/vendor authorizations
- Clause variance
- Implication of change
- Dispute resolution, etc.

The determinants of contract management performance are described below:

i **Risk-sensitive procurement planning** – The CPM must ensure that the procurement planning is carried out strategically (well-established procurement process/method based on the item specifications) considering the requirement of the concerned item as per the project schedule requirement; for instance, in case of long lead items the procurement is based on their delivery timeline and the same must be reflected in the project schedule plan as well. In case of unavailability of any item as mentioned in the specifications of the contract, to avoid the risk of unattainability of the items, alternative materials should be explored by the CPM, such as what, when, and from whom (source) to buy in order to avoid in delay arising due to lack in procurement strategy. The CPM must ensure that the type of delivery model chosen should be efficient, and how the procurement plan should be coordinated and integrated with the project schedule. It also needs to be ensured that the key procurement items should be identified in the early stages with due consideration towards currency and legal jurisdiction (Project Management Institute, 2017a).

106 *Value-driven performance assessment*

ii **Effective planning of contractual obligations** – The CPM must ensure and manage all procurement relationships and should monitor the contract performance of the respective stakeholders and their obligations/requirements to be fulfilled by them. The contractual obligations should be monitored by the CPM to facilitate the implementation of the procurement plan based on which payments can be made as defined in the agreed contract document.

iii **Effective management of contractual obligations** – The CPM must ensure that all contractual obligations related to the project are being met and all stakeholders are aware of their set of responsibilities and obligations as agreed in the legal contract document. The CPM should maintain properly documented records to prevent any disputes which might arise in the future. The whole idea behind proper coordinated and controlled planning is to prevent disagreements between the stakeholders.

iv **Effective claim management** – Due to changes in the scope of work, the unavailability of key items for procurement might lead to some additional costs for which stakeholders might claim for the variation from the agreed scope of work. The claims are also referred as contested changes (Project Management Institute, 2017a). The CPM is responsible for granting necessary approvals, make changes, and corrections as requested by the stakeholders. In case of any claims being made by other parties involved, the CPM must verify the reliability of the claims being put forward, and all claims and their approvals must be formally documented throughout the lifecycle of a project. The CPM should have the capability of identifying, addressing, managing, and resolving any potential claims made by the parties. When these claims cannot be resolved, they become disputes and require alternative dispute resolution techniques to be implemented as defined in the agreed contract between the respective stakeholders.

v **Planning contract closeout** – The project closeout phase involves the finalization of all deliverables as per the scope of work as agreed in the contract document; the engaged resources can be released for new endeavours (Project Management Institute, 2017a). Before final closeout, the CPM must ensure that the project has met its objectives and works are completed including all handover documents, closing of project accounts, formal acceptance by the client, archiving of the necessary information for future use, lessons learned, planning for any excess material, finalization of claims, reallocation of project resources, etc. The project manager should also ensure that completion certificates are received, performing final audits, etc.

Table 23 represents the contract performance determinants as they aggregate in the proportion of their weight towards overall contract performance and their applicability with consideration to PMBoK process groups is represented in Figure 28.

Table 23 Determinants of contract performance

Contract performance	W5	Weightages	Determinants
		W41	Risk-sensitive procurement planning
		W42	Effective planning of contractual obligations
		W43	Effective management of contractual obligations
		W44	Effective claim management
		W45	Planning contract closeout

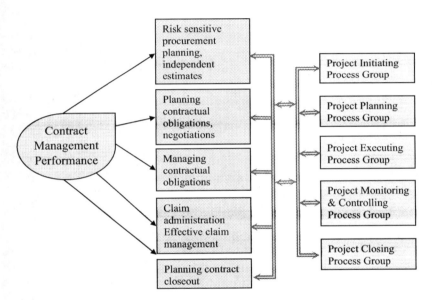

Figure 28 Determinants of contract management performance and PMBOK process groups.

4.2.9 Design management performance indicators and determinants

"Design is a hierarchical activity, defined as a set of plans and a process (how those plans will be achieved)" (Budawara, 2011). The design phase of any construction project has a significant impact on the overall performance and efficiency of a project. Design management performance just simply does not mean the evaluation of the design but also includes the process of design and its effect on the success of the project with respect to the client and organization.

Design performance management needs to be assessed at two distinct levels, the first being the ability to identify missing information or the absence

108 *Value-driven performance assessment*

of certain requirements. At the second level, the logical relationships between the requirements and their functional coordination with the rest of the design need; identification of appropriateness of the fundamental design approach. In this case, the former has a consequence of "syntax error" while the latter can be on account of "logical error", questioning the wisdom of the project team having long-term perpetual functional performance deficiencies.

The measurement of design management performance requires a comprehensive list of indicators that can justify its evaluation process. It should incorporate both the design process as well as the financial attributes as design is a major contributor to any project's cost. Making changes in the early design phase requires the least amount of effort and therefore demands more attention to reduce the overall project costs (Budawara, 2011).

The basis of design as the starting point of the project solution has a critical role in long-term design inappropriateness. The assumptions laid out therein are invariably reasons for the ineffectiveness of design solutions leading to perpetual productivity loss.

A study by the Construction Industry Institute (CII) states that design effectiveness is an important determinant of a project's success. The design process is a complex one, involving numerous factors, knowledge, and constraints, and requiring an efficient team to handle and deliver the final design outputs (*CII- RT008*, 1987).

The use of proficient knowledge is essential to achieve project success, with the design phase being one of the most crucial phases of any project's life cycle. As per *RIBA Plan of Work 2020* (2020), the design team of any project is responsible for designing the project as well as bringing out the information associated with execution. Many specialist consultants – with detailed knowledge and experience of a particular subject – may be involved in the design phase of a project. Out of the seven stages of the RIBA (Royal Institute of British Architects) plan of work, three stages are dedicated to design management itself, i.e. concept design, spatial coordination, and technical design, as represented in Table 24. These three stages

Table 24 Stages of RIBA plan of work 2020

Stage number	RIBA stages
0	Strategic definition
1	Preparation and Brief
2	Concept Design
3	Spatial coordination
4	Technical Design
5	Manufacturing and Construction
6	Handover
7	Use

Source: www.architecture.com

Value-driven performance assessment 109

have a bearing on the comprehensiveness and logical suitability of design invariably specific for each project peculiarity.

Design inputs given right from the project brief stage tend to enhance the effectiveness of the design strategy at an early stage of the project. Inputs to the design can occur at any stage of the project lifecycle right from the concept stage, followed by schematic design, detailed design, tender design, up to the execution stage. Major gathering of input for the design generally occurs during the concept design stage of a project.

4.2.9.1 Determinants of design management performance indicators

The determinants of design management performance indicators to be considered as identified by *CII- RT008* (1987) are mentioned and presented in Table 25.

- Accuracy of Design Documents
- Usability of Design Documents
- Cost of Design
- Constructability
- Economy of Design
- Performance Against Schedule
- Ease of Start-Up
- Security

Some of the key input variables that impact the design effectiveness are:

- Scope Definition
- Owner Profile and Participation
- Project Objectives and Priorities
- Pre-Project Planning
- Basic Design Data
- Designer Qualification and Selection

Table 25 Initial design evaluation criteria

S.NO.	Design evaluation criteria	Qualitative	Subjective
1	Accuracy of design documents		
2	Usability of design documents		
3	Cost of design effort		
4	Constructability of design		
5	Economy of design		
6	Performance against schedule		
7	Ease of start up		

Source: CII- RT008, 1987.

110 *Value-driven performance assessment*

- Project Manager Qualifications
- Construction Input
- Type of Contract
- Equipment Sources

Design is an overall difficult process to manage and involves a large number of personnel such as architects, structural consultant, mechanical, electrical and plumbing consultants, landscape consultant, transportation consultant, sustainability consultant, marketing consultant, etc. This makes the role of a CPM even more complicated as each of the consultants has different protocols of working. During the design process, it is quite common to note that the plan of work for design evolution and development has vulnerability to infuse design changes leading to conflicts in ordinary circumstances. Different disciplines involved in the project have to deal with the responsibilities of adjacent disciplines with understanding (Dr-Ing Schnellenbach-Held & Steiner, n.d.). One way to ensure design coordination in a multi-disciplinary environment is to infuse collaborative working by "process ownership", wherein the leadership is shared depending upon the critical role played by one and the rest supporting the process owner. This also ensures that leadership is not a positional prerogative but a responsibility to take decision and for others to support the lead discipline. In such a situation, the CPM ensures unitary purpose in the overall sense of leadership without infringing on the domain-specific superiority in a particular situation.

The CPM is not expected to be proficient in all disciplines and therefore, must perform design management function as an overarching proactive facilitator rather than getting into the shoes of each discipline leader by himself or herself. Further complications to the performance of a CPM in this respect may arise when the client, being the supreme driver of the project, primarily interfaces with the project manager. In such a situation, the CPM must perform the role of a facilitator without humbling the technical leadership of the design process. The crux, therefore, of the CPM's performance lies in the balance to deal with the domain dominant team in a unified solution-focused process.

A study by Budawara (2011) focuses on the use of summary reports for each design phase, integration of design with supply chain, innovation, and re-use of design experience as some of the major variables to be considered as design performance measures.

The CPM needs to manage the overall process of design and ensure that an outstanding design is delivered keeping in mind the stakeholder expectations. The points of concern with respect to design which are needed to be measured and looked into are stakeholder satisfaction, construction cost and time, the hierarchy of final decision-making power, overall process of design management, etc.

Value-driven performance assessment 111

A study by CRIRA (2004) has highlighted indicators that are needed for the assessment of any design process – integration of design with supply chain, internal time/cost management, risk, reuse of design experience, innovation, client user experience, and stressing on the importance of identifying client needs and integrate it with the design process.

The design performance determinants which impact the overall performance of the project and require better monitoring and control by the project manager are described in the following section:

i **Establishing stakeholder engagement process** – The CPM must coordinate with the stakeholders involved in the project to ensure that the design is in accordance with the design brief and that all consultants on board are able to resolve the design issues. The CPM must establish a design documentation standard for receiving the coordinated drawings and proper procedures to establish design documentation records and revisions like drawing log register, design schedule, design review meetings, etc. to have all the issued drawings and records documented. The project manager only acts as a coordinator or reviewer of the design documents as received, and individual stakeholders can raise their queries through request for Information (RFI) if any issues are identified by him/her in the received drawings and other documents (KOSKELA et al., 2002).

ii **Establishing need-centric design process** – An integrated design-centric process should be established by the project manager to ensure that the final good-for-construction drawings are available for the execution work and that all necessary stakeholders are involved in the design process. The CPM is aware of the need for the design process as per the scope of work and should try to formulate a design-centric strategy by establishing collaborative processes in adherence to standards, legislation, and relevant codes of practice.

iii **Establishing decision-making hierarchy** – The ultimate decision-making power stays with the client and CPM but it requires setting up of standard communication protocol related to design decisions. An integrated design management decision-making programme based on design responsibility matrix should be developed to ease down the process of decision-making and develop the mutual interest of all stakeholders towards the deliverables of the project.

iv **Resolving conflicting interests** – As a CPM, the resolution of conflicts linked with the execution difficulties needs to be identified and communicated to the respective design consultants so as to avoid any conflicts related to design and execution, which might lead to delays in the planned schedules. All design-related problems should be identified and rectified before they begin to manifest as design flaws. In order to involve stakeholders, the best way to resolve conflicts as per the industry

112 *Value-driven performance assessment*

Table 26 Determinants of design performance

Design performance	W4	Weightages	Determinants
		W51	Establishing stakeholder engagement process
		W52	Establishing need centric design process
		W53	Establishing decision-making hierarchy
		W54	Resolving conflicting interests
		W55	Effective planning for scope creep
		W56	Resolving time-cost impacts

practice approach is by developing a mock-up ready as per the approved design drawings and specifications and getting client approvals and any other issues related to design during the execution works can also be easily identified and rectified by the decision making of the stakeholders all together.

v **Effective planning for scope creep** – The determinant of scope creep planning relates to the design changes which require the implementation of an effective change management plan. Since the accumulation of design changes leads to scope creep, to ensure effective design change management, project professionals must periodically review the design and resolve conflicts between the stakeholders to avoid any deviation from the original design scope.

vi **Resolving time-cost impacts** – The determinant of resolving time-cost impacts relates to the value engineering exercise of following a systematic approach to derive the best functional balance between cost, reliability, and performance of the overall design in the lifecycle of the project. Circularity being an emerging concern in climate change, the implications of deconstruction and demolition must be a part of the overall design suitability. It also involves dealing with the design constraints and recommendations for alternative options to reduce its impact on the overall project's time and lifecycle cost (Zimmerman & Hart, 1982).

Table 26 represents the design performance determinants as they aggregate in proportion of their weight towards overall design performance.

4.3 Inferences

The performance indicators of a CPM are broadly defined under five main categories of time, cost, scope, contract, and design. A detailed description of the determinants of each performance indicator has been described as identified through literature and mapped against PMBoK guide based on

Value-driven performance assessment 113

the knowledge areas which will be used as the variables for defining the VDPI in the subsequent chapters of the book. The variables are derived based on project requirements, the skillsets, the knowledge, and competence delivered by the CPM as an individual.

The process of identification of these variables is based on in-depth literature study and expert interview approach conducted in this research. Considering the relation between the project management knowledge areas, performance indicators, and their determinants, these variables of performance are used as input in the equation of VDPI to derive the index score.

References

Abramovici, A. (2000). *Controlling scope creep.* PM Network. https://www.pmi.org/learning/library/controlling-scope-creep-4614

Ahadzie, D. K., Proverbs, D. G., & Olomolaiye, P. O. (2008a). Model for predicting the performance of project managers at the construction phase of mass house building projects. *Journal of Construction Engineering and Management, 134*(8). https://doi.org/10.1061/ASCE0733-93642008134:8618

Ahadzie, D. K., Proverbs, D. G., & Olomolaiye, P. O. (2008b). Critical success criteria for mass house building projects in developing countries. *International Journal of Project Management, 26*(6), 675–687. https://doi.org/10.1016/J.IJPROMAN.2007.09.006

AIPM. (2021). *AIPM professional competency standards for project management-.*

Alarcón, L. F., Grillo, A., Freire, J., & Diethelm, S. (2001). Learning from collaborative benchmarking in the construction industry. In Ninth Annual Conference of the International Group for Lean Construction (IGLC-9).

Ali, H. A. E. M., Al-Sulaihi, I. A., & Al-Gahtani, K. S. (2013). Indicators for measuring performance of building construction companies in Kingdom of Saudi Arabia. *Journal of King Saud University - Engineering Sciences, 25*(2), 125–134. https://doi.org/10.1016/J.JKSUES.2012.03.002

Bannerman, P. L. (2008). *Defining project success a multilevel framework.* https://www.pmi.org/learning/library/defining-project-success-multilevel-framework-7096

Barber, E. (2004). Benchmarking the management of projects: A review of current thinking. *International Journal of Project Management, 22*(4), 301–307. https://doi.org/10.1016/J.IJPROMAN.2003.08.001

BIS. (2009). *IS 15883-1 (2009): Construction project management - Guidelines, Part 1: General.*

BIS. (2013). *IS 15883-2 (2013): Construction project management - Guidelines, Part 2: Time Management.*

Budawara, N. (2011). *Key performance indicators to measure design performance in construction.* Library and Archives Canada = Bibliothèque et Archives Canada.

Chan, A. P. C., & Chan, A. P. L. (2004a). Key performance indicators for measuring construction success. *Benchmarking, 11*(2), 203–221. https://doi.org/10.1108/14635770410532624/FULL/XML

Chan, A. P. C., & Chan, A. P. L. (2004b). Key performance indicators for measuring construction success. *Benchmarking, 11*(2), 203–221. https://doi.org/10.1108/14635770410532624

114 *Value-driven performance assessment*

Chan, A. P. C., Scott, D., & Lam, E. W. M. (2002). Framework of success criteria for design build projects. *Journal of Management in Engineering, 18*(3), 120–128.

Chen, P., Partington, D., & Wang, J. N. (2008). Conceptual determinants of construction project management competence: A Chinese perspective. *International Journal of Project Management, 26*(6), 655–664. https://doi.org/10.1016/j.ijproman.2007.09.002

Cho, K. M., Hong, T. H., & Hyun, C. T. (2009). Effect of project characteristics on project performance in construction projects based on structural equation model. *Expert Systems with Applications, 36*(7), 10461–10470. https://doi.org/10.1016/J.ESWA.2009.01.032

Chou, J. S., & Yang, J. G. (2012). Project management knowledge and effects on construction project outcomes: An empirical study. *Project Management Journal, 43*(5), 47–67. https://doi.org/10.1002/PMJ.21293

CII- RT008. (1987).

CIPS. (2007). *The Charted Institute of Purchasing and Supply (CIPS): Contract Management Guide.* https://www.cips.org/intelligence-hub/member-only/contract-management-contract-mediation-process-guide-part-1

Claessens, B. J. C., Eerde, W. van, Rutte, C. G., & Roe, R. A. (2007). A review of the time management literature. *Personnel Review, 36*(2), 255–276. https://doi.org/10.1108/00483480710726136

Cockfield, R. W. (1987). Scope management | PMI. *PM Network, 1*(3), 12–15. https://www.pmi.org/learning/library/scope-management-9099

CRIRA. (2004). *Benchmarking the performance of design activities in construction (C618).*

Dr-Ing Schnellenbach-Held, U.-P. M., & Steiner, D. (n.d.). *Change management concepts for structural design in early planning phases.*

Garnett, N., & Pickrell, S. (2000). Benchmarking for construction: Theory and practice. *Construction Management and Economics, 18*(1), 55–63. https://doi.org/10.1080/014461900370951

Guide to Contract Management | CIPS. (n.d.). Retrieved April 25, 2022, from https://www.cips.org/knowledge/procurement-topics-and-skills/developing-and-managing-contracts/contract-management/

Harhad Meriem. (2018). PM world contract. *PM World Journal, 7*(10), 1–11.

Jainendrakumar, B. T. (2015). Project scope management in PMBOK made easy. In *PM World Journal Project Scope Management in PMBOK made easy: Vol. IV.* www.pmworldlibrary.net

Kagioglou, M., Cooper, R., & Aouad, G. (2001). Performance management in construction: A conceptual framework. *Construction Management and Economics, 19*(1), 85–95. https://doi.org/10.1080/01446190010003425

Koskela, l., Huovila, P., & leinonen, J. (2002). Design management in building construction: From theory to practice. *Journal of Construction Research, 03*(01), 1–16. https://doi.org/10.1142/S1609945102000035

KPI Report for The Minister for Construction. (1999). http://www.detr.gov.uk

Loosemore, M., & Muslmani, H. S. A. (1999). Construction project management in the Persian Gulf: Inter-cultural communication. *International Journal of Project Management, 17*(2), 95–100. https://doi.org/10.1016/S0263-7863(98)00030-1

Luu, V. T., Kim, S. Y., & Huynh, T. A. (2008a). Improving project management performance of large contractors using benchmarking approach. *International Journal of Project Management, 26*(7), 758–769. https://doi.org/10.1016/j.ijproman.2007.10.002

Value-driven performance assessment 115

Luu, V. T., Kim, S. Y., & Huynh, T. A. (2008b). Improving project management performance of large contractors using benchmarking approach. *International Journal of Project Management*, *26*(7), 758–769. https://doi.org/10.1016/J. IJPROMAN.2007.10.002

Markovic, L., Dutina, V., & Kovacevic, M. (2011). Application of benchmarking method in the construction companies. *Facta Universitatis - Series: Architecture and Civil Engineering*, *9*(2), 301–314. https://doi.org/10.2298/fuace1102301m

Meade, P. (1998). *A guide to benchmarking*. University of Otago.

Mirza, M. N., Pourzolfaghar, Z., & Shahnazari, M. (2013). Significance of scope in project success. *Procedia Technology*, *9*, 722–729. https://doi.org/10.1016/j. protcy.2013.12.080

Mohsini, R. A., & Davidson, C. H. (1992). Determinants of performance in the traditional building process. *Construction Management and Economics*, *10*(4), 343–359. https://doi.org/10.1080/01446199200000030

Nahod, M.-M. (2012). Scope control through managing changes in construction Proje. *Organization, Technology and Management in Construction an International Journal*, 438–447. http://www.grad.hr/otmcj/clanci/vol4_is1/OTMCJ_2012_4_1_ web_clanak_9.pdf

Nassar, N. K. (2009, October). An integrated framework for evaluation of performance of construction projects. *PMI® Global Congress*. https://www.pmi.org/ learning/library/evaluation-performance-construction-projects-6751

Oliveros, J., & Vaz-Serra, P. (2018). Construction project manager skills: A systematic literature review. *52nd International Conference of the Architectural Science Association*, 185–192.

Orihuela, P., Pacheco, S., & Orihuela, J. (2017). Proposal of performance indicators for the design of housing projects. *Procedia Engineering*, *196*, 498–505. https://doi. org/10.1016/j.proeng.2017.07.230

Pariafsai, F., & Behzadan, A. H. (2021). Core competencies for construction project management: Literature review and content analysis. *Journal of Civil Engineering Education*, *147*(4), 04021010. https://doi.org/10.1061/(asce)ei.2643-9115.0000051

Park, S. H. (2009). Whole life performance assessment: Critical success factors. *Journal of Construction Engineering and Management*, *135*(11), 1146–1161. https:// doi.org/10.1061/(ASCE)CO.1943-7862.0000090

Paul, V. K., & Basu, C. (2021). *A handbook for construction project planning and scheduling*. COPAL Publishing Group. https://books.google.co.in/books?id=4e-ecDwAAQBAJ&printsec=frontcover&source=gbs_ge_summary_r&-cad=0#v=onepage&q&f=false

Project Management Institute. (2005). *Practice standard for earned value management*.

Project Management Institute. (2017). *A guide to Project Management Body of Knowledge (PMBOK guide): Vol. Sixth edition*.

Reichel, C. W. (2006). Earned value management systems (EVMS). *PMI® Global Congress*. https://www.pmi.org/learning/library/earned-value-management-systems-analysis-8026

RIBA Plan of Work 2020. (2020). www.ribaplanofwork.com

Skibniewski, M. J., & Ghosh, S. (2009). Determination of key performance indicators with enterprise resource planning systems in engineering construction firms. *Journal of Construction Engineering and Management*, *135*(10), 965–978. https:// doi.org/10.1061/(ASCE)0733-9364(2009)135:10(965)

116 *Value-driven performance assessment*

Takim, R., & Akintoye, A. (2002). Performance indicators for successful construction project performance. In *University of Northumbria. Association of Researchers in Construction Management* (Vol. 2).

Toor, S. ur R., & Ogunlana, S. O. (2010). Beyond the "iron triangle": Stakeholder perception of key performance indicators (KPIs) for large-scale public sector development projects. *International Journal of Project Management, 28*(3), 228–236. https://doi.org/10.1016/J.IJPROMAN.2009.05.005

Unegbu, H. C. O., Yawas, D. S., & Dan-asabe, B. (2022). An investigation of the relationship between project performance measures and project management practices of construction projects for the construction industry in Nigeria. *Journal of King Saud University - Engineering Sciences, 34*(4), 240–249. https://doi.org/10.1016/J.JKSUES.2020.10.001

Vyas Gayatri, S., & Kulkarni Saurabh, S. (2013). Performance Indicators for Construction Project. *International Journal of Advanced Electrical and Electronics Engineering, (IJAEEE), 2*(1), 61–66.

Walker, D. H. T., & Rowlinson, S. M. (2008). *Procurement systems : A cross-industry project management perspective.* 455.

What Is a Key Performance Indicator (KPI)? (n.d.). Retrieved April 18, 2022, from https://kpi.org/KPI-Basics

Yang, H., Yeung, J. F. Y., Chan, A. P. C., Chiang, Y. H., & Chan, D. W. M. (2010). A critical review of performance measurement in construction. *Journal of Facilities Management, 8*(4), 269–284. https://doi.org/10.1108/14725961011078981

Zimmerman, L. W., & Hart, G. D. (1982). *Value engineering: A practical approach for owners, designers, and contractors.* Van Nostrand Reinhold.

5 Threshold performance level for time management performance

5.1 Introduction

This chapter defines the criteria for defining five levels of threshold performance assessment of a CPM based on the ISO: 9004:2008 & 2018 code for quality management. Further, in other sections of the chapter, the levels for assessing time management performance for a CPM are described against each determinant of the time management performance indicator. Section 5.3 and subsequent sections describe each level of performance on which CPM's performance can be gauged at individual and organizational levels.

5.2 Criterion for defining five levels of threshold performance

For defining the threshold performance levels of the VDPI concept, the International Organization for Standardization (ISO) code 9004:2000 & 2018 versions are referred through which the concept of establishing five levels of maturity for assessing performance has been adapted. The concept of maturity levels as defined in the Quality Management System Guidelines for Performance Improvement ISO:9004: 2000 (2000) are based on the intent of assessing and improving the effectiveness of the organizational management, by measuring the performance progress against the set objectives. Though the ISO: 9004:2000 is for quality management, the essence of capturing the self-assessment maturity model in the standard finds its wider applicability across various organizations. The threshold performance levels derived in the VDPI concept are also based on a self-assessment approach by an individual or an organization, which are important in continuous assessment of the performance with respect to the project goals.

The self-assessment application in VDPI tool provides one with an opportunity to compare with the organizational goals and define one's own parameters for improvement, to ensure that the continuous development process of an individual doesn't get hindered and best possible efforts and practices can be identified.

Similar to ISO:9004 a scale range of levels 1 to 5 as represented in Table 27 has been used in VDPI assessment criterion, its intent is to provide a simple and user-friendly methodology to determine the level of performance

DOI: 10.1201/9781003322771-5

118   *Threshold performance level for time management performance*

Table 27 Performance levels

S. No.	Performance level	Level No.
1	No formal approach	Level 1
2	Reactive approach	Level 2
3	Stable formal system approach	Level 3
4	Continual improvement approach	Level 4
5	Best in class performance	Level 5

Source: Quality Management System Guidelines for Performance Improvement ISO:9004: 2000, 2000.

of an individual construction project manager based on the project success parameters also referred to as project objectives. These levels 1–5 are being referred to as threshold performance levels, which are nothing but a representation of performance competence of an individual considered acceptable by an organization or implemented standard. These levels define the delivered work/results through which the performance of the individual/ organization can be easily inferred.

The level range of 1–5 represents no formal systematic approach to best-in-class performance. The identified threshold performance levels can be applied in determining the next level of maturity needed to thrive for better performance by offering a comprehensive quantifiable analysis management process based on self-assessment.

Based on the evaluation of VDPI score for an individual as demonstrated by the delivered work in the project, different levels of performance can be defined for the individual. The finalized VDPI scores can lead to multiple performance levels. Configuration of threshold performance levels might also vary from organization to organization, so the described threshold performance levels in the current as well as the subsequent chapters are not mandatory levels as they might vary from organization to organization.

5.3 Classification of time performance threshold levels

The processes that a CPM should follow to determine his/her performance is illustrated in levels 1–5 (Table 28). Each level specifies the standard of performance an individual must achieve when carrying out a function in the workplace, together with the knowledge and understanding they need to meet the desired objectives. The criteria in Level 5 define "Best Practice/ Processes" which gradually decreases for each level. The interpretation required for an individual during performance evaluation based on the criteria mentioned in Table 28 is explained in the subsequent section.

5.3.1 Planning work coordination (W_{11})

Planning is the first step in developing the best course of action to accomplish clearly defined objectives. Planning is a rational, dynamic, and integrative

Threshold performance level for time management performance 119

process, not only limited to determining the time performance but also responsible to manage other performance indicators i.e., cost, scope, contract, and design. As multiple stakeholders are involved in any construction project thus, for effective planning the CPM needs to establish coordination amongst them.

- **Level 1** – Determining project objectives is the first step in planning. For CPM, it is important to determine the project objectives and prepare a project schedule to get an idea about what needs to be done, which resources must be utilized, and when the project is due? But for the less complex projects, many times the execution takes place in absence of a formal project schedule. This doesn't indicate that the CPM had not planned the work at all, rather he did the planning using his wisdom but did not translate the same formally in the form of a project schedule.

    The CPM who has planned the activities but has not prepared a formal project schedule falls under Level 1.
- **Level 2** – The CPM is expected to have a master schedule considering major milestones with specific deadlines but not necessarily prepared a detailed schedule of all work packages separately. However, the decision behind the level of detailed schedule depends upon the complexity and scale of the project, for a small and less complex project even a master schedule or milestone schedule would suffice the purpose.
- **Level 3** – Often on construction sites, despite having project schedules the execution happened in isolation without following the planned project schedule. This may happen due to various reasons like unreasonable schedules, the schedules prepared not taking consensus with the stakeholders, etc. All this led to a situation of project delays, cost overruns, disputes, etc. Thus, a CPM needs to prepare a project schedule involving all the key stakeholders.
- The CPM who has the detailed schedules but the schedules are not coordinated or agreeable to the key stakeholders lies at this level.
- **Level 4** – The CPM should ensure the availability of a coordinated project schedule at the site. Though establishing coordination and taking consensus with all stakeholders is quite challenging but the CPM should establish a formal process of coordination or involve key stakeholders while developing detailed schedules. The CPM must ensure that everyone is working toward the same identified objectives.
- **Level 5** – This level indicates the "Best Practices" where it is expected that CPM has the coordinated project schedules for respective work packages. To effectively manage the challenges of coordination and to ensure smooth coordination amongst key stakeholders CPM should deploy advanced techniques like BIM, AI, etc. but the decision of using such advanced techniques must be taken considering the complexity and scale of the project.

120 *Threshold performance level for time management performance*

5.3.2 *Effective schedule control (W_{12})*

Effective schedule control is the most important aspect to determine whether the project is behind or ahead of planned deadlines. Ensuring the adherence to work progress as per the planned schedule and taking necessary actions to bring back the progress of work as per the planned schedule is an important aspect to avoid the issues of Time and Cost Overruns in construction projects.

- **Level 1** – At the lowest level, it is not possible to formally monitor the work progress in absence of any formal project schedule, however, at this level, the CPM should apply his wisdom and experience to take appropriate measures required for controlling the progress of work.
- **Level 2** – Often, CPM is unable to follow the planned schedule during execution. Mostly the reason is the irregularity in project schedule updation. While interviewing CPMs, they revealed that the execution started as per schedule but in case of any missed deadline they don't have an established process to update the schedule regularly as a result the execution deviates from the originally planned schedule. Thus, the CPM who has master/milestone schedules in place but doesn't update the schedule regularly to ensure work progress in alignment with originally planned deadlines falls under this level.
- **Level 3** – Irregular project schedule updation leads to deviations and if these deviations are not identified in the beginning, it starts accumulating and result in uncontrolled. Uncontrolled deviations impacted significantly the project duration and budget. If a CPM has a schedule in place and implemented schedule tracking and monitoring processes but yet unable to control project deviations due to irregular project updation lies under this level.
- **Level 4** – This is the desired level, where a CPM is expected to have all schedule tracking, monitoring, and control processes in place. At this level, the CPM should anticipate any schedule deviation and take desired interventions to minimize its impact.
- **Level 5** – The best processes that aid in effective schedule control includes continuous project tracking and monitoring based on set milestones, updating the project processes, understanding the impact in terms of delay, taking desired interventions to make the project progress back on the planned schedule, and deploying advanced techniques i.e., BIM, AI, etc. for real-time monitoring and updating the project schedule. Regular monitoring and control measures are essential as it keeps the schedule deviation in control.

5.3.3 *Risk forecasting (W_{13})*

Due to the complexity of construction projects, merely the development of the project schedule won't give a realistic idea of project duration. Thus,

Threshold performance level for time management performance 121

a CPM needs to anticipate and incorporate the uncertainties so that an appropriate risk response is developed and the project's level of exposure is controlled. Risk forecasting involves prioritizing risks and assessing each identified risk's probability of occurrence and potential impact. To acquire a true depiction of these risks – and opportunities – throughout the project, CPM should implement risk management measures and incorporate the results into the forecasting process.

- **Level 1** – At the lowest level, the CPM should anticipate the major risks and take necessary measures to minimize their impact on project duration but due to the absence of a formal project schedule at this level, the CPM is not expected to implement a formal process of risk forecasting.
- **Level 2** – At this level, the CPM should anticipate the risks using his wisdom and experience. However, no formal risk allocation or risk loading is expected in the project schedule. It is expected that the project schedule developed by CPM is deterministic without translating the anticipated risks while arriving at the project duration.
- **Level 3** – At this intermediate level, if the CPM can translate anticipated risk in the form of buffer (additional duration) against activities then he lies in this level. Though probabilistic scheduling techniques provide better results but that process in itself is cumbersome and required historical data. The applicability of probabilistic scheduling also depends upon the precision, scale, and typology of the project. Hence, it is expected that the CPM should anticipate risks, translate them into a form of buffer and incorporate them into the project schedule.
- **Level 4** – Identifying all possible risks is the most crucial step for risk forecasting. But this step is best performed by the project team rather than by one individual. Thus, the CPM should engage the key stakeholders to strengthen the process of risk identification. After identifying risks, the CPM should ensure these risks are well incorporated into the project schedule and accordingly prepare the risk response to control the project exposure.
- **Level 5** – Being the highest level, it is expected that the formal processes of risk identification, risk loading in project schedules, and risk response are well in place. The CPM at this level should be a step ahead to forecast risks by applying their knowledge and experience and taking appropriate measures to minimize the impact of risks on project time and cost.

5.3.4 *Effective resource planning (W_{14})*

CPM must utilize its resources in the best possible way. Maximizing the productivity of resources is the key aspect to achieving the desired goal by efficiently managing them. The CPM should ensure to align their resources to the right tasks and set realistic expectations. In addition, the CPM should

122   *Threshold performance level for time management performance*

apply his wisdom and experience proactively to resolve the crisis. Resource planning is quite a complex process and inefficient resource planning may be impacted both time and cost, so the CPM should be focused on simplifying the complex process of resource planning and optimizing resources to achieve the set project objectives.

- **Level 1** – In absence of a formal project schedule, the formal resource plan/schedule is not expected from the CPM at this level. However, the CPM should be able to quantify the resource requirement throughout the project cycle using his experience and wisdom and allocate them accordingly based on their skill set and project requirement.
- **Level 2** – Many times CPM does develop separate resource schedules based on the work packages and scope but those resource schedules are not in coherence with the project schedule resulting in the issues like allocating a greater number of resources than desired, allocating wrong resources for the wrong tasks, resource crisis, and not achieving the desired productivity. However, despite non-coherence between resource schedule and project schedule, the CPM at this level is expected to use his wisdom and experience to manage their resources to meet the desired project objectives.
- **Level 3** – This is the intermediate level where CPM should have a project schedule with loaded resources against the activities. The number, types, skill sets, capability, and duration of resources for which they are required at the site should be clearly defined at this level. In addition, the CPM should have a system in place for regular resource monitoring for progress, efficiency, and effectiveness in delivering their expected project contribution.
- **Level 4** – This is the desired level for any construction project. It is expected that the CPM should have a resource-loaded resource schedule and ensure that the resources are allocated and utilized efficiently. CPM should have established a process to optimize resources and minimize idle manhours to maximize productivity. At this level, CPM should attempt to automate the processes and manual tasks. The CPM should identify any mundane tasks and remove them to avoid errors and improve productivity.
- **Level 5** – This is the highest level, where in addition to the criteria mentioned in Level 4, the CPM should deploy the latest tools and technologies for resource planning. Tools like BIM, AI, Robotics, etc. shall be quite useful to improve efficiency. The focus would be on minimizing waste and duplication, streamlining and automating processes, and maximizing and speeding throughput.

5.3.5 *Controlling delays (W_{15})*

In any construction project delays are inevitable. Thus, the challenge for any CPM is how well he takes measures to control delays balancing the

Threshold performance level for time management performance 123

project cost. CPM must understand how project delays creep up while managing projects. Regularly tracking project schedules, anticipating critical causes of delays, and taking appropriate measures to minimize the impact are a few ways to deal with them when they inadvertently occur. Cancelling delays often requires additional costs in the form of additional resources, increased working hours, enhanced machinery, advanced technology, etc., thus the CPM is expected to trade off the additional cost for cancelling delays or executing a delayed contract till its end. Managing delays also depends upon the complexity of the construction project as the complexity brings in more uncertainty resulting in delays. Hence, the level of competence, processes, and tools required by the CPM to control delays often depends upon the scale, typology, complexity, and nature of the construction project.

- **Level 1** – In absence of a formal project schedule, one cannot monitor and strategize to control delay. Stagewise milestones tracking, monitoring, and control are not expected at this level. But the CPM should be proactive in its approach to anticipate the probable cause of delays and take appropriate measures to minimize the impact of delays for achieving the project objectives.
- **Level 2** – At this level, the CPM should apply their experience to anticipate the reason for the delay, identify the critical activities and take appropriate measures to minimize the impact of delays. The CPM should ensure managing cost overruns that happened due to delays in the project timeline. The deployment of scientific tracking, monitoring and control processes like EVM, EVS, etc., by the CPM, is not expected at this level.
- **Level 3** – This is the intermediate level where the CPM should regularly track and monitor the delay using scientific tools like EVM, EVS, etc. The CPM is expected to take account to see whether the progress of work is on target. The CPM should perform regular checks to identify the variations (if any) either in the project schedule or budgeted cost. In case of any observed variance, the CPM should actively communicate the same to key stakeholders seeking early resolution.
- **Level 4** – In addition to the criteria mentioned in Level 4, the CPM should take appropriate actions in consultation with key stakeholders to minimize the impact of delay. The application of scientific tools must be extended to forecast the impact of delay at a given point and extrapolate to predict the result at the end of the project. The CPM is expected to take proactive measures for course correction in case there is any delay at a given point to finish the project achieving set objectives.
- **Level 5** – This is the highest level where the CPM is expected to have real-time tracking and monitoring process in place. The application of advanced tools i.e., BIM, AI, RFID, etc. is expected at this level. Though,

Table 28 Threshold performance levels for time management performance indicator determinants

S.no	Performance indicators	Weightage of indicators (equal weightage to be moderated as per organizational discretion if any)	Determinants of performance (identified from literature, expert interviews, and field studies; weightages (Wm) derived from questionnaires survey as perception of the field experts)		Performance level (Level m) Levels 1 to 5 represent performance of a specific Project Manager in a given project situation. As level 1 being the lowest and level 5 being the highest having numerical weight varying from 1 to 5. Based on the actions undertaken by the project manager performance level is assigned.				
		(Pn)	(Wn)	m Wm Determinants	Level 1	Level 2	Level 3	Level 4	Level 5
1	Time Performance	W1		1 W11 Planning work coordination	Non-availability of project schedule	Master schedule is available but the detailed schedule of all work packages is not available.	Master schedule and detailed schedule are available without the involvement of stakeholders.	Master and detail schedule is available in consensus all the stakeholders.	Coordinated schedule duly approved by stakeholders is available and advance techniques viz. BIM, AI and laser technologies are deployed by the CPM
				2 W12 Effective schedule control	Supervision of work progress in absence of project schedule	Non-adherence of the master schedule to execute the work at site	Work progress as per schedule but schedule deviation is uncontrolled	Continues project monitoring and tracking is in place to adhere with planned scheduled timelines	Project monitoring and control on real time basis using advanced techniques viz. BIM, AI, etc.

| 3 | W13 | Risk forecasting | No provision of risk loading in absence of project schedule | Deterministic schedule without risk loading | Deterministic schedule with additional buffer incorporating risks without realizing the real project conditions | Preparation of comprehensive schedule loaded with all possible risks considering project conditions | Application of advanced tool such as Montecarlo simulation, Fuzzy logic, AI, etc. to forecast risk and load them to update the schedule consistently. |
| 4 | W14 | Effective resource planning | No provision of resource planning in absence of project schedule | Resource schedule available but not in coherence with project schedule. | Project schedule loaded with resources but underutilized/ unutilized in real time | Project schedule is loaded with deployed resources at site ensuring planned productivity | Application of advanced tools to minimize idle manhours and optimize resources to ensure timely completion of project within the stipulated budget. |

(Continued)

S.no	Performance indicators	Weightage of indicators (equal weightage to be moderated as per organizational discretion if any)	Determinants of performance (identified from literature, expert interviews, and field studies; weightages (Wm) derived from questionnaires survey as perception of the field experts)	Performance level (Level m) Levels 1 to 5 represent performance of a specific Project Manager in a given project situation. As level 1 being the lowest and level 5 being the highest having numerical weight varying from 1 to 5. Based on the actions undertaken by the project manager performance level is assigned.				
			5 W15 Controlling delays	In absence of project schedule, the stagewise milestone control is deficient	Controlling delays based on experience without applying scientific tools and techniques such as EVM, EVS, etc.	Application of scientific tools and techniques such as EVM, EVS, etc. to track the delay.	Application of scientific tools and techniques such as EVM, EVS, etc. to identify the extent of delay and appropriate actions to be taken accordingly to overcome the impact of delay.	Regular tracking and monitoring of project schedule using advanced techniques and application of project management techniques to optimize project delay with respect to project cost on real time.

the decision to apply such advanced tools depends upon the complexity of the project. Data collection is crucial at this level which helps CPM to forecast any deviations till the end of the project. CPM should efficiently utilize its resources for real-time tracking and monitoring and should communicate any change in plan amongst key stakeholders.

5.4 Inferences

The criteria for defining the threshold performance levels (1–5) of the determinants related to time management performance have been detailed based on which an individual CPM can assess his/her performance against the time management performance indicator. Using this, they can also identify their threshold level of performance by evaluating oneself based on the description of each performance level for each determinant of time management performance.

Reference

Quality Management System Guidelines for Performance Improvement ISO:9004: 2000, (2000).

6 Threshold performance level for cost management performance

6.1 Introduction

This chapter defines the criteria for defining five levels of threshold performance assessment of a CPM. It includes the description of processes that a CPM must follow to determine his/her cost performance. These processes are described against each determinant of cost management performance indicator for the threshold levels of performance. Section 6.2 and subsequent sections describe each level of performance against each determinant, based on which CPM's performance can be gauged at individual and organizational levels

6.2 Classification of cost performance threshold levels

The processes that a CPM should follow to determine his/her cost performance are illustrated in Levels 1–5 (Table 29). Each level specifies the standard of performance an individual must achieve when carrying out a function in the workplace, together with the knowledge and understanding they need to meet the desired objectives. The criteria in Level 5 define "Best Practice/Processes" which gradually decreases for each level. The interpretation required for an individual during performance evaluation based on the criteria mentioned in Table 29 is explained in the subsequent section.

The project details and project attributes that are in any-way related directly to the project cost or any ample quantity of the amount anyway belongs to the project activity can be categorized under this domain.

6.2.1 Effective cash flow management (W_{21})

This is the most critical determinant determining the success of any construction project. Having plenty of liquid cash allows CPM to invest in resources, increase productivity, and manage expenses comfortably for a smooth cash flow. The accurate estimation of cash flow in the early stages of a project is considered a vital factor that indicates the project's financial significance. Cash flow planning is a crucial step in making significant

DOI: 10.1201/9781003322771-6

Threshold performance level for cost management performance   129

decisions concerning how to liquidate a project with cash. Though, the competence that CPM has to effectively manage the cash flow depends upon the scale, complexity, and nature of the project. Levels 1 to 5 below explain the criteria/skill set of a CPM in detail:

- **Level 1** – At this lowest level, the CPM is expected to have a cash flow statement but not in detail for the separate work packages. Formally or informally every CPM does have some cash flow calculations to get an idea about the requirement of cash at major stages of the project. However, at this level, the formal cash flow statements aligned with the project schedule are not expected. But the CPM must have a clear idea about the cash requirement, probable expenses, and sources to arrange finances to achieve major milestones.
- **Level 2** – It has been observed while speaking with various CPMs that mostly they do prepare cash flow statements and project schedules but are often not aligned with each other. The main reason behind this is the lack of coordination and the absence of an information-sharing mechanism. As a result of unaligned cash flow statements with the project schedule, the real application of cash flow statements to anticipate the cash requirement at various stages of the project would not suffice for the purpose. At this level, the CPM is expected to take proactive measures to manage cash flows including invoicing customers promptly, offloading unutilized or underutilized resources, and closely monitoring the expenses.
- **Level 3** – At this level, the CPM is expected to deploy all necessary processes for cash flow management. However, despite having formal cash flow management processes in place, often the project failed due to the deviation of work progress from the original schedule, lack of monitoring of expenses, and deviation in scope. To effectively manage the cash flow, the CPM is expected to ensure whether the actual progress of work and expenses follow the originally estimated budget. In case of any deviation, the CPM should take necessary measures to cut off unnecessary expenses without impacting the project objectives.
- **Level 4** – In addition to the formal cash flow management processes in place, if a CPM deployed a process to regularly track and monitor the physical work progress, identify deviation (if any), anticipate the impact of deviation for complete project duration, and take necessary measures to reduce the impact in terms of time and cost, lies under this level. The CPM is expected to maintain the cash liquidity to accomplish the planned activities, ensure timely invoicing and bill payments, and communicate with key stakeholders to resolve any crisis well in advance.
- **Level 5** – Real-time tracking and monitoring the physical work progress is a complex and cumbersome process. Thus, at this highest level, the CPM should use advanced tools for forecasting and monitoring cash flows. Data collection is crucial at this stage, required to perform

130 *Threshold performance level for cost management performance*

extrapolation during forecasting. CPM is expected to anticipate any crisis based on the past performance of the project and realign his strategy in consultation with key stakeholders.

6.2.2 Controlling budget variance (D_{22})

This is an important determinant to check periodically whether the progress is on track to meet the project objectives or not. Preparing a budget and sticking to it are two different challenges that CPM has to manage. At the initiation stage itself, the CPM should ensure preparing a detailed budget considering the detailed analysis of each factor. If the budgeting process is not realistic or detailed, then analyzing budget variance would not make any sense. Checking periodically for any variation is key to figuring out whether everything is going according to plan and whether any corrective actions are necessary. Because the budget acts as a blueprint for carrying out the project objectives, variance analysis assists CPM in determining whether or not the targets will be met and what efforts should be made to guide the project back on track. However, the skill set and competencies of CPM to control budget variance depend upon the nature, complexity, typology, and scale of the project. The criteria defining skill and competencies set for a CPM are categorized and explained in Levels 1 to 5 in a subsequent section.

- **Level 1** – This is the lowest level of skill set where due to the absence of a detailed budget the CPM is not expected to deploy formal processes to identify the budget variance. Indeed, the CPM should apply experience and wisdom to track project progress periodically for any possible variation in the project budget. Accordingly, the CPM should take proactive actions to further eliminate the impact of budget variance resulting in project cost overruns.
- **Level 2** – At this level, the CPM is expected to make adjustments in the budget itself based on the results from the budget variance analysis. The CPM should be competent enough to compare the actual versus planned progress to identify the variation in budget. However, the absence of a detailed schedule implying formal budget variance control processes aligned with the project schedule is not expected at this level.
- **Level 3** – In addition to the criteria mentioned in Level 2, the CPM should have formal budget variance control processes in place. The periodic tracking and monitoring processes to compare actual work progress versus planned work should be implemented and CPM should regularly update the schedule to realistically forecast the project variance.
- **Level 4** – The CPM should ensure accurate and timely accounting to create a database required to forecast budget variance till project completion. The CPM should establish major milestones and conduct periodic checks to analyze how well the work progress is sticking to the originally planned budget. The CPM should be competent to compare

Threshold performance level for cost management performance 131

the actual results to the budget values for the same period to analyze the variances. Accordingly, the CPM should take appropriate corrective actions to achieve the budget targets. Consequently, at this level, it is essential for CPM to regularly update the forecast based on the information gleaned from the variance analysis and the courses of action taken to complete the project within set objectives.

- **Level 5** – This level exhibits "best in class" practices where in addition to the criteria mentioned in Level 4, the CPM should integrate the budget forecasting with real-time data using advanced tools such as BIM, Navisworks, AI, etc. The interactive dashboards to display real-time data are expected at this level which streamlines the coordination through sharing real-time information with key stakeholders.

6.2.3 Managing risk contingencies (D₂₃)

A contingency plan is an important tool for managing risks that ensures the viability of construction projects in case of any uncertainty. As we discussed earlier, the budget is prepared based on assumptions, as the execution starts the projects tends to encounter many deviations from the original assumptions considered. Thus, contingency refers to a percentage of money reserved to cover unanticipated costs not identified or assumed during budget preparation. In general practice, the contingency percentage ranges from 5% to 10% of the total construction budget. The contingency cost is reserved and not allocated to any specific activity or area of work. This determinant is indirectly linked with other determinants including controlling delays, cost overruns, scope creep, and managing changes. However, the performance of CPM depends on how well he used this amount as a backup to maintain project deadlines and quality commitments established for the project.

- **Level 1** – In absence of a formal project schedule, resources scheduling, and cash flow statements, the CPM is not expected to anticipate risk factors. Thus, the development of a formal risk contingent plan by the CPM is not expected at this level. The usage of contingency amounts completely lies in the wisdom of CPM to achieve the project objectives. However, the CPM should use the contingent amount as the last resort for overcoming any deviation.
- **Level 2** – Since the CPM is expected to have a milestone schedule thus, the CPM should apply his experience and knowledge to anticipate the risk factors resulting in cost overruns. However, the impact of anticipated risk factors in terms of the amount of cost overrun throughout the various stages of the construction project cycle is a complex task and not expected from the CPM at this level. CPM should ensure to not utilize the contingent amount considering it as a right, rather the focus should be on anticipating and managing risk well in advance to eliminate the requirement of using contingent amount.

132 *Threshold performance level for cost management performance*

- **Level 3** – At an intermediate level, the CPM is expected to deploy the risk identification and risk quantification processes. Quantifying the impact of anticipated risk factors is a complicated task and requires data collection. At this level, the CPM is not expected to have a mechanism for data collection in place. But based on experience and wisdom CPM should anticipate the impact of risk in terms of cost overruns and communicate the same to key stakeholders for early resolution.
- **Level 4** – This is the desired level where in addition to the criteria mentioned in Level 4, the CPM should establish a process to collect data required to forecast the impact of risk in terms of cost overrun. The formal contingent plan with identified risk is expected at this level. Accordingly, the CPM should communicate the forecasted risk impact to stakeholders and take corrective actions in consultation with stakeholders to achieve the project objectives.
- **Level 5** – Anticipating and quantifying risks is the most essential and complex task for which the CPM should deploy advanced tools for real-time data collection and use that data to extrapolate the risk impact in terms of cost overruns. Accordingly, CPM is expected to prepare a contingency management plan and ensure its integration with change management processes. In case of delay or cost overrun or both, the CPM should judiciously utilize the available contingency amount to overcome the impact in consultation with stakeholders.

6.2.4 *Controlling cost overruns (D$_{24}$)*

The determinants, controlling cost overruns, budget variance, and risk contingencies are interrelated with each other. All three have a similar objective to complete the project scope within a stipulated budget. Controlling cost overrun is an exhausting task that requires constantly tracking the project costs and ensuring things are going as planned at all stages. In addition to this, CPM needs to monitor other project constraints i.e., time, scope, risk, and quality. Detailed and realistic estimation, controlling changes owing to managing scope, having a risk response plan, and establishing coordination processes are major measures that can reduce the extent of cost overruns. To effectively control project cost overruns, a CPM should identify the causes of overruns and delays and act proactively to take corrective actions in the earlier phases of the project. As the uncertainties increases based on the project complexity so would the possibility of cost overrun. Thus, for effective management different skill set of CPM is required depending upon the nature and complexity of a project.

- **Level 1** – Managing cost overruns at this level is quite difficult for CPM in absence of a formal project schedule, risk management plan, and cash flow statements. However, for smaller projects, CPM can still manage overruns due to less probability of uncertainties. If, the CPM is unable to anticipate the possible causes of cost overrun, unable to forecast the

Table 29 Threshold performance levels for cost management performance indicator determinants

S.no	Performance indicators	Weightage of indicators (equal weightage to be moderated as per organizational discretion if any)	Determinants of performance (identified from literature, expert, interviews and field studies; weightages (Wm) derived from questionnaires survey as perception of the field experts)			Performance level (Level m) Levels 1–5 represent performance of a specific Project Manager in a given project situation. As Level 1 being the lowest and Level 5 being the highest having numerical weight varying from 1 to 5. Based on the actions undertaken by the Project Manager performance level is assigned.				
	(Pn)	(Wn)	m	Wm	Determinants	Level 1	Level 2	Level 3	Level 4	Level 5
2	Cost Performance	W2	1	W21	Effective cash flow management	Non-availability of details of cash flow with respect to work packages as per project schedule	Availability of cash flow as per work packages but not aligned with schedule	Cash flow details are available with respect to project schedule but not confirming actual work done	Cash flow is available for entire work packages and regularly monitored and updated with respect to actual work progress	Application of advanced tool such as BIM, AI, etc. for forecasting, and monitoring of cash flow with respect to real time work progress.
			2	W22	Controlling budget variance	Identification of budget variance purely on the basis of experience	Identification of budget variance limited to a certain interval of completion of milestones	Identification of budget variance for all the milestones and activities limited to planned scheduled completion	Identification of budget variance for all the milestones and activities as per actual work progress and take appropriate measures to control ensuring the project completion within stipulated budget	Application of advanced tools such as BIM, Navisworks, AI, etc. to provide real-time budget variance s per actual work progress and provide appropriate measures to control ensuring the project completion within stipulated budget

(Continued)

S.no	Performance indicators	Weightage of indicators (equal weightage to be moderated as per organizational discretion if any)	Determinants of performance (identified from literature, expert, interviews and field studies; weightages (Wm) derived from questionnaires survey as perception of the field experts)			Performance level (Level m) Levels 1–5 represent performance of a specific Project Manager in a given project situation. As Level 1 being the lowest and Level 5 being the highest having numerical weight varying from 1 to 5. Based on the actions undertaken by the Project Manager performance level is assigned.				
	(Pn)	(Wn)	m	Wm	Determinants	Level 1	Level 2	Level 3	Level 4	Level 5
			3	W23	Managing risk contingencies	Inability to anticipate the risk factors resulting in cost overrun of project activities and utilizing the available contingency amount to balance the cost overrun.	Ability to identify the risk factors responsible for the cost overrun but inability to quantify the impact of risk in terms of cost.	Ability to identify and quantify the impact of risk factors in terms of cost	Ability to identify and quantify the impact of risk factors in terms of cost and take appropriate measures to reduce the impact of risk resulting additional cost	Application of advanced tool to forecast risks as per project activities; and identify and quantify the overall impact in terms of cost. To provide strategies to control the cost overrun through utilizing the available contingency amount optimum.
			4	W24	Controlling cost overruns	Inability to identify the cost overrun during the project execution	Ability to identify the cost overruns for major milestones and taking control measures-based experience	Ability to identify the cost overruns for major milestones and taking control measures-based experience	Ability to identify the cost overruns for each activity, anticipate the possible extent of cost overruns for remaining activities and taking preventive actions to control them to minimize/ eliminate the cost overrun.	Application of advanced tool to track and identify the cost overrun in real time basis and take appropriate action to control cost overruns

Threshold performance level for cost management performance   135

impact, and waits for remedial action to minimize the impact of cost overrun in the later stages of the project falls under this level.

- **Level 2** – Identification of the probable causes of cost overrun is the first and most crucial step. To effectively manage cost overruns, the CPM should identify the causes in the early phases of the project. However, at this level, it is difficult for CPM to quantify the impact of risks owing to cost overruns till project completion in absence of a detailed project schedule and risk response plan. Though, the CPM is expected to communicate the possibility of cost overruns amongst stakeholders and team members on regular basis to seek early remedial actions.
- **Level 3** – At this level, the CPM should periodically track the work progress, extrapolate the same in terms of overall project completion, and make a comparison between the actual work and planned work to identify deviation. Accordingly, the CPM is expected to prepare an action plan in consultation with project stakeholders to manage cost overruns and complete the project to achieve set objectives.
- **Level 4** – A good CPM must be capable of identifying the possible sources of cost overruns and mitigating their effect at the early stages of the project itself as its impact keep accumulating as the project progress. Regular tracking is the most robust tool to identify any deviation. In addition, for forecasting the extent of impact CPM should ensure to deploy data collection processes. At this level, the CPM should inform the key stakeholders periodically regarding the change in the action plan and take preventive measures in coherence with stakeholders to minimize the impact of cost overruns.
- **Level 5** – At this highest level, the CPM should deploy advanced tools for real-time tracking, monitoring, and data collection of actual work progress. In addition to the criteria mentioned in Level 4, the CPM is expected to implement thorough project planning processes, stick to the planned scope, keep stakeholders informed about the change in management strategy, and engage stakeholders using real-time dashboards and other such tools to strengthen coordination. Any observed deviation from the originally planned schedule must be immediately communicated to the key stakeholders to develop an action plan for early resolution.

6.3 Inferences

The criteria for defining the threshold performance levels (1–5) of the determinants related to cost management performance have been covered in detail, based on which an individual CPM can assess his/her performance against the cost management performance indicators related to any project's success and identify their threshold level of performance by evaluating oneself based on the description of the performance levels against each determinant of cost management performance.

7 Threshold performance level for scope management performance

7.1 Introduction

This chapter defines the processes that a CPM should follow to determine his/her scope management performance as described in threshold performance Levels (1–5). Each level specifies the standard of performance an individual must achieve when carrying out a function in the workplace, together with the knowledge and understanding they need to meet the desired objectives. Section 7.2 of the chapter provides a description of the performance levels (1–5) that have been described for each determinant of scope management performance.

7.2 Classification of scope performance threshold levels

The processes that a CPM should follow to determine his/her scope management performance is illustrated in Levels 1–5 (Table 30). Each level specifies the standard of performance an individual must achieve when carrying out a function in the workplace, together with the knowledge and understanding they need to meet the desired objectives. The criteria in Level 5 define "Best Practice/Processes" which gradually decreases for each level. The interpretation required for an individual during performance evaluation based on the criteria mentioned in (Table 30) is explained in the subsequent section.

7.2.1 Coordinated scope planning (W_{31})

Finalizing project scope is one of the most critical attributes of project management. Since multiple stakeholders having varied responsibilities are involved in construction projects and it is interesting to note that quite often these stakeholders work in isolation from each other. In simple words, the scope and responsibility of one stakeholder are unknown to the other, and so forth. For example, the contractor is unaware of the scope of work of the design consultant. Similarly, the civil contractor is unaware of the scope of work of an interior contractor. Though mostly it is not preferable and

DOI: 10.1201/9781003322771-7

Threshold performance level for scope management performance   137

required to share the complete scope of work of every stakeholder amongst the project team, thus a CPM must share the key details of the scope of work amongst the project team so that every team member can able to discharge their assigned responsibility with better clarity. It is evident from the studied construction projects as well as through literature that poor coordination may lead to the issues of rework, scope creep, and changes that ultimately lead to delay and cost overruns. Thus, CPM needs to ensure coordination amongst various stakeholders and team members to successfully achieve the project objectives.

- **Level 1** – At this lowest level, the CPM is not expected to establish a formal engagement process between stakeholders. As a result, the projects encountered frequent scope changes resulting in time and cost overruns. However, if the CPM ensures to have final scope provided by clients and monitors the work progress to periodically check for any scope deviation lies at this level.
- **Level 2** – Here, the CPM is expected to finalize the scope by taking consensus from key stakeholders. This reduces the probability of frequent changes in scope, although even in case of any changes in scope during execution the CPM should identify the stakeholder responsible for scope changes and communicate the same to the client. In addition, the CPM should update the project schedule periodically based on the revised scope. However, due to the absence of a detailed project schedule at this level, it is expected that the execution of work may not follow the revised schedule and may deviate from the scope.
- **Level 3** – Establishing engagement processes between stakeholders, developing a project scope plan involving key stakeholders, and addressing the stakeholder's needs throughout the project life cycle are quite complex challenges for CPM. At this level, the CPM should ensure to communicate the project scope plan with project team members to ensure that the work is not progressing in isolation from each other. However, the CPM is not expected to periodically update the scope incorporating changes and accordingly update the project schedule, resources, cash flows, and risk response plans.
- **Level 4** – In addition to the criteria mentioned in Level 3, the CPM should regularly incorporate changes, quantify their impacts in terms of time and cost, and update the project scope in consultation with stakeholders. CPM must communicate the revised scope to the associate project team members and ensure whether the work is progressing following the revised scope.
- **Level 5** – Establishing a communication channel between stakeholders and project team members is essential for CPM. Thus, at this highest level, the CPM is expected to implement advanced tools like MIS, ERP, Dash Boards, etc. to share real-time information amongst team members. Implementation of such advanced IT tools also quickens the

138　*Threshold performance level for scope management performance*

approval process and avoids errors. In addition, the CPM should fulfil the criteria mentioned in Level 4.

7.2.2 Effective stakeholder involvement (W_{32})

In continuation with the previous determinant, the stakeholder involvement process encompasses consideration of different opinions and interests of the stakeholders and addressing them throughout the project life cycle. The CPM should take the responsibility to involve the right people in the right way by developing a stakeholder engagement plan. This process can mitigate potential risks and conflicts with stakeholder groups. It is evident in various construction projects that poor stakeholder involvement led to unclear project scope resulting in uncontrolled changes and uncontrolled increment in project scope. A good project manager should ensure to finalize the scope addressing the client's requirement to avoid frequent changes during the project life cycle.

- **Level 1** – Involving key stakeholders, understanding their requirements, and translating them to define the project scope is the three steps a CPM should perform. However, at this lowest level, the CPM is not expected to define the project scope. Indeed, this level is suitable for small projects having a single client. In such projects, the scope is generally decided by the client and communicated to CPM. But the CPM is expected to clearly understand the scope, prepare an execution plan, anticipate uncertainties in the form of changes, and communicate the same to the client at the early stages of the project itself.
- **Level 2** – At this level, the CPM must prepare a stakeholder's responsibility matrix required to establish a communication process. The CPM should involve key stakeholders i.e., consultants and clients to finalize the scope. However, the intervention of consultants and CPM is very limited in this process of scope finalization. This skill set of CPM is suitable for small and less complex projects where the number of stakeholders involved is less. The CPM is expected to regularly coordinate with the consultant and client for early resolution of changes.
- **Level 3** – This level is applicable for medium-scale projects having multiple clients. Establishing a clear communication plan is essential at this level. In addition, the CPM must ensure the involvement of key stakeholders in finalizing the scope. The CPM should communicate the finalized scope to project team members, perform consultative meetings, and update the project scope with key stakeholders addressing team member concerns. CPM should monitor the progress of work periodically to ensure whether the work is progressing within a specified scope. The CPM should act as a coordinating channel between project team members, and consultants to address and clarify the issues raised by the project team ensuring timely delivery of the project.

Threshold performance level for scope management performance 139

- **Level 4** – This is the desired level applicable to medium to large-scale and complex projects. In addition to the criteria mentioned in Level 3, the CPM should have a system in place to regularly track and monitor changes. In case of any changes, the CPM must act proactively to resolve them through consultative meetings with stakeholders. However, if the change is unavoidable then CPM should anticipate the impact of the change on the project scope and communicate the same to the client. The CPM is expected to seek an early resolution from the client and update the scope accordingly.
- **Level 5** – It may be noted that sharing real-time work progress and the latest action plan is the biggest challenge for any CPM to effectively manage the scope changes. Thus, at this highest level, the CPM is expected to implement advanced IT tools like MIS, ERP, Dashboards, etc. to create an effective communication process. Furthermore, the CPM should ensure that the user interface used to share information is acceptable to project team members and stakeholders.

7.2.3 *Monitoring project deliverables* (W_{33})

Effective scope monitoring starts with a robust plan that defines the project scope along with work packages and tasks needed to achieve the project objectives. While preparing a detailed scope plan, a CPM ensure to include all deliverables along with a clear process to achieve them. The timelines and cost of each task and assigning the responsibility of the project team members are key to completing the project. The CPM must communicate clear expectations throughout the project to stay on track and makes monitoring effective. However, CPM should adjust the expectations regularly based on the information gathered throughout the monitoring process. In case the actual work progress doesn't align with the planned progress, the CPM must evaluate the situation and revise the strategy accordingly. The challenge for CPM is to stay on track and ensure the work progress is closer to the planned scope. However, in case of any deviation, the revised scope plan with updated timelines and budget must be prepared in consensus with key stakeholders. Often, CPM finds it difficult to decide the frequency of project monitoring. Constant monitoring or expected daily reporting from the project team develops a feeling of distrust between the team members. Similarly, increasing the frequency of project monitoring led to a situation where a CPM finds out too late the project has deviated from the original scope. Thus, the CPM must decide the frequency of monitoring considering the scale, nature, and timeline of the project.

- **Level 1** – Breaking down the project scope into various work packages and assigning the responsibilities is the first step of scope monitoring. Thus, if the CPM is having clarity about project scope, derived work packages and tasks defining complete scope, and assigned

140 *Threshold performance level for scope management performance*

responsibilities and timelines accordingly, lies under Level 1. At this lowest level, the CPM is not expected to regularly monitor the scope formally as per the scope plan. But the CPM is expected to collect data to find out any deviation in scope using his experience and wisdom.

- **Level 2** – In addition to the preparation of the scope plan, the CPM is expected to ensure whether the actual work is progressing as per the planned scope. The CPM must collect the data on regular basis through daily or weekly reporting of work progress by project team members. The CPM should try to identify the deviation (if any) at the early stage itself to minimize the impact of scope deviation on overall project objectives. Accordingly, the CPM should revise his strategy using his experience and wisdom to stay on track and complete the project by meeting planned objectives.
- **Level 3** – The performance of the CPM is assessed by how well he managed the project in a crisis. The deviation may happen despite having a detailed scope plan and micromanagement of work packages. But the challenge for CPM is how well and quickly he revised his strategy to take the scope back on track. At this level, the CPM is expected to convey any deviation immediately to the key stakeholders. Devise strategy to manage deviation minimizing the impact on project timelines and budget in coherence with project stakeholders.
- **Level 4** – Often disputes arise in construction projects due to the deviation in scope. So, in addition to the criteria mentioned in Level 3, the CPM should take the due approval from key stakeholders in case of revisions in the project scope. Many times, the extent of deviation is unmanageable and forces decision-makers to revise the project scope. In such a situation, the CPM should prepare the revised scope in consultation with stakeholders and ensures to take formal approval.
- **Level 5** – Identification of deviation is the first and most critical task in the scope monitoring process. Thus, at this highest level, the CPM is expected to deploy advanced and automated tools to gather real-time data and simplify the project monitoring process. The CPM should update the scope plan regularly using the collected data to ensure the work progress is in alignment with the scope plan. In addition, the CPM should implement ERP/MIS system to maintain smooth communication between stakeholders and project team members.

7.2.4 Controlling scope creep (W_{34})

The fourth and last determinant defining scope performance is controlling scope creep. Scope creep is defined as the uncontrolled changes in the project resulting in deviation in project scope that was originally agreed upon. The continuous or uncontrolled growth in the project scope may be because of the poor definition of the project scope in the early stage, poor control of the project during the execution, or maybe because of mismanagement

Threshold performance level for scope management performance 141

or improper documentation. Changes are inevitable in any construction project and every CPM do face moments where they need to reassess the originally planned strategy and revise it accordingly. A good CPM should be adaptable and know how to manage, negotiate, and handle when the project is moving away from the original plan. Managing changes right from the inception stage is essential for any CPM otherwise these small changes cumulatively add up throughout various project stages and ultimately turn into big and uncontrolled at a later project phase. Thus, CPM must have a scope of work in the initial stage itself and keep the work progress on track through robust monitoring and control.

- **Level 1** – Generally scope creep happens when the new requirements are kept added by stakeholders even after the project execution has started. Often these changes are not reviewed properly that resulting in increased project scope. Many times, the CPM takes due approval of the additional requirements and revised the scope plan accordingly yet faced major challenges while managing the project timelines and budget if these changes happen at the later stages of the project. Thus, to effectively control these changes the CPM should develop a scope management plan and deploy strategies to control changes. However, at this lowest level, the CPM is not expected to formally prepare and monitor work progress using scope management and change management plans.
- **Level 2** – At this level, the CPM is expected to prepare a detailed scope management plan describing how the scope of the project will be established and controlled. In addition, the scope management plan must include the approval process by the stakeholders in case of any changes in the original project scope. The CPM is expected to regularly monitor and control the work progress as per the scope management plan, communicate immediately to the stakeholders for any changes, and develop a revised scope plan in coherence with all stakeholders and team members to complete the project objectives.
- **Level 3** – At this intermediate level, the CPM should deploy scope management and change management plan to control project scope. In addition, the CPM is expected to regularly track and monitor the work progress, compare it with the original plan, and identify the deviations (if any). The CPM should anticipate the impact of these deviations on project costs and timelines. If the deviation is unmanageable, the CPM should take a lead and discuss with stakeholders how the change fits into the overall project and devise a revised strategy to meet the project objectives.
- **Level 4** – In addition to the criteria mentioned in Level 3, the CPM should quantify the impact of scope changes, adjust the project timelines, and budget and communicate the impact to project stakeholders. Fixing responsibilities is an essential step to effectively managing changes. Often, scope creep happens due to design changes made by

Table 30 Threshold performance levels for scope management performance indicator determinants

S.no	Performance indicators	Weightage of indicators (equal weightage to be moderated as per organizational discretion if any)	Determinants of performance (identified from literature, expert interviews, and field studies; weightages (Wm) derived from questionnaires survey as perception of the field experts)		Performance level (Level m) Levels 1 to 5 represent performance of a specific Project Manager in a given project situation. As Level 1 being the lowest and level 5 being the highest having numerical weight varying from 1 to 5. Based on the actions undertaken by the Project Manager performance level is assigned.				
	(Pn)	(Wn)	m	Wm	Determinants	Level 1	Level 2	Level 3	Level 4	Level 5
3	Scope Performance	W3	1	W31	Coordinating scope planning	No coordination mechanism between all the stakeholders involved in the project	Project scope planning is in consensus of all the stakeholders but the execution is in isolation	Project scope planning is coordinated with all the stakeholders and project execution done as per the approved scope plan.	Project scope planning is coordinated with all the stakeholders and project execution done as per the approved scope plan and any changes in scope throughout the project are approved by the stakeholders.	Application of advanced tools like MIS, ERP etc. to provide common interface between all the stakeholders from where they have the information about scope changes
			2	W32	Effective stakeholder involvement	Involvement of all the stakeholders is deficient during the scope finalization	Limited involvement of stakeholders (client, consultant) during the finalization of scope	Involvement of all the stakeholders during the finalization and execution of project scope.	Involvement of all the stakeholders during the finalization and execution of project scope. Any changes in scope of project are duly consulted and approved between the stakeholders.	Application of advanced tools like MIS, ERP etc. to provide common interface between all the stakeholders to expedite the decision on scope changes.

3	W33	Monitoring project deliverables	Preparation of schedule of deliverables as per project scope	Preparation of schedule of deliverables as per project scope and ensuring the work progress as per project schedule	Preparation of schedule of deliverables as per project scope and ensuring the work progress as per project schedule. Any changes in the project deliverables are intimated to the concerned stakeholders.	Preparation of schedule of deliverables as per project scope and ensuring the work progress as per project schedule. Any changes in the project deliverables and scope approved by the concerned stakeholders.	Advanced tools to identify the scope variance in real time basis and taking approval from the stakeholders
4	W34	Controlling scope creep	Formal change control procedure is deficient	Identifying regular project scope measurement but no scope variance analysis	Identifying regular project scope measurement along with scope variance analysis	Identifying project scope managing changes to control the scope variance to limit the time and cost overrun	Application of advanced tools such as BIM, Navisworks, etc. to identify project scope managing changes to control the scope variance to limit the time and cost overrun.

144 *Threshold performance level for scope management performance*

consultants, additional requirements by clients, rework, wrong execution by contractors, etc. so the CPM should regularly document these changes and fix the responsibilities of the respective team member and stakeholders. At this level, implementing a formal change control process is expected from the CPM.

- **Level 5** – At this highest level, the CPM is expected to implement advanced tools for real-time monitoring and controlling changes. Tools like BIM, Navisworks, etc. would be quite handy for real-time monitoring and effective coordination. In addition, establishing a clearer communication process is equally important where the advanced IT tools like ERP, MIS, etc. would be quite effective. The CPM must ensure that the real-time data is being collected and shared between project team members and stakeholders.

7.3 Inferences

The criteria for defining the threshold performance Levels (1–5) of the determinants related to scope management performance have been detailed in the chapter, based on which an individual CPM can assess his/her performance against the scope management performance indicator and identify their threshold level of performance by evaluating oneself as per the description of each performance level and the processes followed against each determinant of scope management performance.

8 Threshold performance level for contract management performance

8.1 Introduction

This chapter defines the processes that a CPM should follow to determine his/her contract management performance as described in threshold performance Levels 1–5. Each level specifies the standard of performance an individual must achieve when carrying out a function in the workplace, together with the knowledge and understanding they need to meet the desired objectives. In Section 8.2, Levels (1–5) have been described for each determinant of contract management performance.

8.2 Classification of contract management performance threshold levels

The processes that a CPM should follow to determine his/her design management performance is illustrated in Levels 1–5 (Table 31). Each level specifies the standard of performance an individual must achieve when carrying out a function in the workplace, together with the knowledge and understanding they need to meet the desired objectives. The criteria in Level 5 define "Best Practice/Processes" which gradually decreases for each level. The interpretation required for an individual during performance evaluation based on the criteria mentioned in (Table 31) is explained in the subsequent section.

8.2.1 Risk-sensitive procurement planning (W_{41})

Setting up a standardized procurement process to maintain reliable relationships with vendors, suppliers, team members, and stakeholders is key to successful project delivery. The CPM should identify the materials with lead time, ordering quantities, and procurement schedule and ensure their compliance with the project schedule. Ultimately the CPM should consider the risks at every stage and employ effective forecasting methods to manage them from the beginning. Failing to manage procurement risks effectively led to non-compliance with legal requirements resulting in huge

DOI: 10.1201/9781003322771-8

146   *Threshold performance level for contract management performance*

fines and penalties. Furthermore, the procurement plan must be developed incorporating the forecasted risks at each stage of the project. For successful project delivery, the procurement process of the project must be regularly reviewed from the procurement planning stage through contract administration. The objective of the review is to learn from what worked and what did not work during the procurement processes.

- **Level 1** – Poor procurement planning could be hazardous to the project. Whether it's the matter of procuring material or an agency, the CPM must know the deadlines of requirements and also the quantity. Procuring too early and too late creates issues of storage, unutilized resources, delays, and cost overruns. Thus, any CPM needs to prepare a comprehensive procurement plan aligned with project timelines. However, often in the case of small projects, the CPM doesn't prepare detailed comprehensive plans, yet manages the project timelines and budget in control using their experience and wisdom.
- **Level 2** – At this level, the CPM should have a formal procurement plan loaded with forecasted risk during various stages of the project. The CPM must have a system to collect data for risk forecasting and prepare risk loaded procurement schedule aligned with project timelines and budget. Though, the CPM is not expected to quantify the impact of these risks in terms of time and cost overruns.
- **Level 3** – In addition to having a formal risk-loaded procurement plan, the CPM should have analytical skills to quantify the impact of risks in terms of time and cost. The CPM must consider the risks attached at every stage, from unreliable vendors to late deliveries. The CPM must focus to standardize the procurement process, so that team members understand the protocols they need to follow, identify delays in the system, and trace back accountability.
- **Level 4** – Ultimately for successful project delivery, managing procurement risks effectively to manage project timelines and budget is essential. Thus, the CPM should employ effective forecasting methods for risk identification from the beginning of the project itself. At this level, in addition to the criteria mentioned in the above three levels, the CPM should act proactively to take necessary reasons for minimizing the impact of risks. For example, in case of non-availability of specified material, the CPM should anticipate it early and have consultative meetings with stakeholders to either change or suggest any alternative specification rather than waiting for the situation to arrive and then react.
- **Level 5** – Often, without sophisticated forecasting and procurement planning, the CPM struggles to understand when and what to order. Thus, analyzing past needs to anticipate future requirements is an essential step in forecasting that requires technological grounding.

Threshold performance level for contract management performance 147

The manual procurement procedures require a lot of manhours for collecting, inputting, and updating data and documentation. Automating the process of purchasing orders and procurement approvals could be very effective for managing project timelines. At this level, the CPM is expected to implement advanced technology that helps a CPM to forecast and monitor each stage of the procurement process.

8.2.2 Planning contractual obligations (W_{42})

The duties that each party is bound by law to carry out under a contract are known as contractual obligations. Additionally, contractual responsibilities are extensive, difficult to handle, and frequently the most overlooked risk management instrument. Contracting parties must satisfy all of their commitments for contracts to perform as promised. Thus, a CPM should regularly track the compliance of the contractual obligations to reduce the project risks.

- **Level 1** – The identification of stakeholders with their key responsibilities relevant to the project delivery is the first step a CPM should take. Considering the complexity of construction projects where the stakeholders are sitting in various locations, yet involved in the project. The CPM who has the capability of identifying stakeholders and mapping their professional liabilities may be awarded Level 1 in this determinant category.
- **Level 2** – Establishing a communication channel is important to bring the stakeholders to a common platform. This unfolds the possibility of working collaboratively where each stakeholder keeps informed about their obligations be it time, cost, and scope. The CPM who has established communication channels between stakeholders falls under this level.
- **Level 3** – In addition to the communication channels, the CPM must deploy the process of reviewing contractual obligations regularly. The real-time information regarding obligations monitoring must be shared and discussed with stakeholders for building a common consensus for evolving solutions and mitigation actions.
- **Level 4** – In addition to the criteria mentioned in Level 3, the CPM should have a rigorous document control process in place. Subsequently, the CPM is expected to prepare a contingency plan and protocols for risks impacting contractual obligations.
- **Level 5** – Effective application of the above-mentioned processes to manage contractual obligations in case of default must be an overarching aim of the CPM. A good CPM is how well he handled and manages the risks and non-compliance of contractual obligations in consensus with the stakeholders involved.

148 *Threshold performance level for contract management performance*

8.2.3 *Managing contractual obligations (W_{43})*

Listing obligations from contracts allows for effectively managing project objectives and monitoring whether the involved agencies are living up to their obligations. If a contracting party does not perform according to the agreement, they are usually obliged to compensate the other party. Thus, establishing an overview of contractual obligations helps CPM to clarify the responsibilities of the different stakeholders which aids CPM to manage conflicts and control the contractual obligations.

- **Level 1** – Understanding contractual obligations for the involved agencies is the first step. Once CPM developed a clear understanding regularly reviewing the obligations to manage them effectively must be the next step. In case of any non-compliance, the CPM should act proactively to undertake suitable actions for managing defaults.
- **Level 2** – At this level, the CPM should establish and manage the communication channels between stakeholders and project team members to monitor the contractual obligations. The CPM must include the deadlines in the contractual obligations implementation plan and should regularly update the document according to the new contracts or contract amendments.
- **Level 3** – The CPM should manage compliance with contracts, policies, processes, and other obligations, as it turns out to be a good risk minimization practice. But the management requires rigorous documentation and regular updation of obligations of involved agencies. At this, the CPM is expected to take contingent actions in cognizance of key stakeholders to minimize the impact of defaults.
- **Level 4** – In addition to the criteria mentioned in the above three levels, the CPM should plan and initiate actions to manage time, cost, and scope change based on non-compliance with contractual obligations.
- **Level 5** – At this highest level, the CPM is expected to employ modern technologies such as Artificial Intelligence (AI) for insights and workflow to enable collaboration. Collaboration can be a very effective tool to strengthen and streamline obligations management. Furthermore, the CPM must create a working environment where each party deals truthfully and fairly with the other party and refrains from fulfilling their part of the agreement by using force or coercion.

8.2.4 *Effective claim management (W_{44})*

Claim management is an inevitable process in construction project management, to reach successfully the desired results. The process needs efficient and effective management during the entire life cycle of a project. Generally, claims relating to the encountered conditions occur during the construction phase. However, the contract document and the information provided or

Threshold performance level for contract management performance 149

withheld during the pre-contract phase are the major factors behind the development of claims. The CPM should try to avoid a claim by managing the causes with due diligence, aligning the paperwork, and ultimately completing it in a timely and professional manner.

- **Level 1** – At the lowest level, the CPM is not expected to have a formal process of claim management. However, the CPM should track the contractual obligations regularly and avoid the situation of claims before they occur.
- **Level 2** – Identifying the deviations and additional work from the project scope is a critical task that every CPM should follow. Untimely resolution of these deviations may be led to claims in long run. The CPM who has employed the formal prevention or mitigation processes of claim management can lie at this level.
- **Level 3** – In construction projects, the term "Claims" and "Change" are quite confusing. What distinguishes a claim from a change is the element of disagreement between the parties as to what is due or whether or not anything is due. If an agreement is reached, then the claim disappears and becomes a change. If not, the claim may proceed to negotiation, mediation, arbitration, and finally, litigation before it is ultimately resolved. For effective management of the claim, the CPM should create an engagement or dialogue between stakeholders to get to a common agreement. Creating an environment of regular dialogue and communication reduces the possibility of claims.
- **Level 4** – To successfully manage claims, the CPM should establish a standardized process before any incident ever occurs. The processes must include the collection of pertinent information regularly and work collaboratively to get the team inputs quickly and correctly for early resolution. Successful claims management processes are built upon a standardized, consistent way of managing claims, and that process is in place before any incident ever occurs.
- **Level 5** – In addition to the criteria mentioned in the above four levels, the CPM is expected to employ advanced tools to identify the additional work or rework on a real-time basis. They must have a system to record these changes, notify the stakeholders in case of any deviation, and take due approvals to avoid any discrepancies in the approval of claims.

8.2.5 Planning contract closeout (W_{45})

A contract closeout occurs when a contract has met all the terms of a contract and all administrative actions have been completed, all disputes settled, and final payment has been made. To successfully close out a contract, the details of the contract may need to be reviewed to ensure that the obligations of the contract were met as expected. The formal closure process

Table 31 Threshold performance level of contract performance indicator

S.no	Performance indicators	Weightage of indicators (equal weightage to be moderated as per organizational discretion if any)	Determinants of performance (identified from literature, expert interviews and field studies; weightages (Wm) derived from questionnaires survey as perception of the field experts)			Performance level (Level m) Levels 1–5 represent performance of a specific Project Manager in a given project situation. As Level 1 being the lowest and level 5 being the highest having numerical weight varying from 1 to 5. Based on the actions undertaken by the Project Manager performance level is assigned.				
	(Pn)	*(Wn)*	*m*	*Wm*	Determinants	*Level 1*	*Level 2*	*Level 3*	*Level 4*	*Level 5*
4	Contract	W4	1	W41	Risk-sensitive procurement planning	Inability to anticipate risk factors resulting poor procurement planning of different work packages to avoid time and cost overrun of the project	Ability to anticipate the risk factors during procurement planning responsible for the time and cost overrun but the inability to quantify the impact of risk.	Ability to anticipate and quantify the impact of risk factors during procurement planning in terms of time and cost	Ability to anticipate and quantify the impact of risk factors in terms of time and cost and take appropriate measures to reduce the impact of risk resulting in additional time and cost of project	Establishing procurement plan consistent with project schedule. Application of advanced tool to forecast risks during procurement planning as per project activities; and identify and quantify the overall impact in terms of time and cost. Early resolution of claims and prevention of disputes. Facilitating expeditious dispute resolution. Monitoring Contractual liabilities consistent with project execution.

| 2 | W42 | Planning contractual obligations | Establish matrix of obligations of stakeholders. Identify organizational interfaces for decision making in accordance with contractual obligations. | Establish communication channels for contractual obligation monitoring. Identifying time, cost, and scope impact. | Establish matrix of obligations of stakeholders and consensus-based obligations review mechanism. Establish communication channels for contractual obligation monitoring and proactive mitigation actions. Rigorous document control for contractual communication. Building consensus for time, cost, and scope impact for evolving solutions. | Establish matrix of obligations of stakeholders and consensus-based obligations review mechanism. Establish communication channels for contractual obligation monitoring and proactive mitigation actions. Rigorous document control for contractual communication. Establishing contingent plan and protocols for risks impacting contractual obligations. Building consensus for time, cost, and scope impact for evolving solutions during currency of the work. | Establish matrix of obligations of stakeholders and consensus-based obligations review mechanism. Establish communication channels for contractual obligation monitoring and proactive mitigation actions. Rigorous document control for contractual communication. Establishing contingent plan and protocols for risks impacting contractual obligations. Building consensus for time, cost, and scope impact for evolving solutions during currency of the work. Planning mechanism for obligation default identification. |

(*Continued*)

S.no	Performance indicators	Weightage of indicators (equal weightage to be moderated as per organizational discretion if any)	Determinants of performance (identified from literature, expert interviews and field studies; weightages (Wm) derived from questionnaires survey as perception of the field experts)			Performance level (Level m) Levels 1–5 represent performance of a specific Project Manager in a given project situation. As Level 1 being the lowest and level 5 being the highest having numerical weight varying from 1 to 5. Based on the actions undertaken by the Project Manager performance level is assigned.				
	(Pn)	(Wn)	m	Wm	Determinants	Level 1	Level 2	Level 3	Level 4	Level 5
			3	W43	Managing contractual obligations	Managing and reviewing obligations and undertaking actions for managing defaults. Managing organizational interfaces for decision making in accordance with contractual obligations.	Managing communication channels for contractual obligation monitoring. Managing time, cost, and scope impact.	Managing and reviewing obligations and undertaking actions for managing defaults. Managing rigorous document control for contractual communication. Undertake contingent actions on risks impacting obligations.	Managing and reviewing obligations and undertaking actions for managing defaults. Managing rigorous document control for contractual communication. Undertake contingent actions on risks impacting obligations. Initiate actions for time, cost and scope change based on contractual implications as well as project utilization.	Managing and reviewing obligations and undertaking actions for managing defaults. Managing rigorous document control for contractual communication. Undertake contingent actions on risks impacting obligations. Initiate actions for time, cost and scope change based on contractual implications as well as project utilization. Record occurrences of events during currency of projects for claim prevention as well as early dispute resolution.

4	W44	Effective claim management	No process of claim management in place in the project	Identification of deviation/ additional work in the project scope resulting in claim during the various stages of project	Identification of deviation/ additional work in the project scope and take intimate the associated stakeholders on regular basis	Identification of deviation/additional work in the project scope and take due approvals from the associated stakeholders to avoid any discrepancies in approval of claims	Applications of advanced tools to identification the deviation/additional work due to rework, scope change, etc. On real time basis and take due approvals from the associated stakeholders to avoid any discrepancies in approval of claims
5	W45	Planning contract closeout	Absence of formal documentation required for contract closeout for the work packages	Formal documentation is in place which is required for contract closeout for the work packages but not effectively implemented	Formal documentation is in place which is required for contract closeout for the work packages and effectively implemented without client consensus.	Effective management of contract closeout adopting the scientific process of contract management and closing out all the work-package confirming to client satisfaction.	Application of advance tools for effective management of contract closeout adopting the scientific process of contract management in real time and closing out all the work-package confirming to client satisfaction.

154 *Threshold performance level for contract management performance*

may vary according to the size and typology of the project. The requirements for contract closeout should be documented within the contract by the CPM.

- **Level 1** – At the lowest level, the CPM is not expected to formally document the contract closeout. However, CPM should ensure that all the obligations are met as per the terms and conditions mentioned in the contract document.
- **Level 2** – At this level, the CPM should formally document the contract closeout ensuring the executed work is acceptable and meets the requirements of the contracts. The CPM must ensure that the scope of work for all the work packages has been completed and is acceptable by the client.
- **Level 3** – In addition to formally documenting the contract closeout, the CPM ensures to complete the snag list in acceptable terms and seeks a formal written notice mentioning the contract closeout from the client. The notice ensures that the executed work is acceptable and that the contract is considered closed.
- **Level 4** – The verification of the executed work by the CPM, the client, key stakeholders, and in some instances the consultants is important to confirm that the contract has been completed in all respect. The CPM must formally document the process of snagging and desnagging to ensure that the final verification is complete and as per the desired satisfaction of the client.
- **Level 5** – Due to the involvement of multiple agencies, the CPM must ensure the verification of completed work by the respective agency at each stage. The CPM is expected to deploy advanced technologies to collect real-time information for monitoring the quality of work and accordingly rectify the defects (if any) during various stages of the project confirming contractual obligations and client satisfaction.

8.3 Inferences

The criteria for defining the threshold performance Levels (1–5) of the determinants related to contract management performance have been detailed out based on which an individual CPM can assess his/her performance against the contract management performance indicator and identify their threshold level of performance by evaluating oneself based on the description of each performance level for each determinant of contract management performance.

9 Threshold performance level for design management performance

9.1 Introduction

This chapter defines the processes that a CPM should follow to determine his/her design management performance as illustrated in Levels 1–5. Each level specifies the standard of performance an individual must achieve when carrying out a function in the workplace, together with the knowledge and understanding they need to meet the desired objectives. In Section 9.2 of the chapter, levels (1–5) have been described for each determinant of design management performance.

9.2 Classification of design performance threshold levels

The processes that a CPM should follow to determine his/her design management performance is illustrated in Levels 1–5 (Table 32). Each level specifies the standard of performance an individual must achieve when carrying out a function in the workplace, together with the knowledge and understanding they need to meet the desired objectives. The criteria in Level 5 define "Best Practice/Processes" which gradually decreases for each level. The interpretation required for an individual during performance evaluation based on the criteria mentioned in Table 32 is explained in the subsequent section.

Design is the process of creating a solution following a project brief and then preparing a plan of action to execute the design. To safeguard the interest of stakeholders, satisfy the project budget, and project timelines, and executed coordinated designs the CPM needs to perform thorough planning. Often, CPM faces issues while executing designs due to lack of information, poorly coordinated drawings, inconsistencies in different sets of drawings, etc. These issues increase based on the project typology, scale, complexity, and involvement of multiple consultants and stakeholders. Thus, a CPM must implement design management processes to manage design throughout the project lifecycle.

9.2.1 Establishing stakeholder engagement processes (D_{51})

This is the first determinant defining design performance. The CPM should engage the stakeholders in the design process with consistent and

DOI: 10.1201/9781003322771-9

156 *Threshold performance level for design management performance*

timely information sharing. Engaging stakeholders right from the design development stage is key as engagement ensures a collaborative process that enables sharing of requirements in a commonplace, and deriving consensus to translate the shared information in a form of design. Often the challenge for CPM is to understand that all stakeholders are not the same but each brings some different information essential for a project's success. Thus a CPM needs to ensure the way of engaging stakeholders based on their capability and how to use the information to strengthen the design process. Ideally, the stakeholder engagement plan is recommended that allows the project manager to devise a systematic approach to ensure expectations, decisions, risk/issues, and project progress information is delivered to the right person at the right time. As the design is an iterative process that requires inputs at each stage and accordingly creates an output thus identifying stakeholders and mapping them based on their capabilities is essential for a CPM. However, often the scale, typology, and complexity of the project guide how detailed a stakeholder's management plan would be? The criteria defining skill sets and competencies that a CPM should have? are enumerated in Levels 1 to 5 as below.

- **Level 1** – As multiple stakeholders are involved in any construction project. Thus, at this crudest level identifying all the stakeholders involved in the design process is the first task that a CPM should do. A comprehensive list of stakeholders along with their responsibilities and capabilities would be a good starting point. Though, a formal stakeholders engagement plan based on the stages in the design process is not expected at this stage. But the CPM must prepare a clear communication plan containing who should be contacted in what situation to ensure effective coordination.
- **Level 2** – Developing a stakeholder engagement plan is the next step after identifying the stakeholders involved in the design process. To encourage collaborative design practices, the CPM should implement the processes of getting inputs and feedback from various stakeholder groups. The stakeholder engagement plan should provide a clear outline of when and how to communicate with stakeholders. The CPM is expected to use a stakeholder engagement plan to share the relevant information amongst stakeholders and manage design changes. So, the CPM who has developed and implemented a stakeholder engagement plan in the design process falls under this level.
- **Level 3** – Often the stakeholder engagement plan is followed only till the design development stage. That means once the final design is approved the consultants start working in isolation and here the actual problem starts. As design development is an iterative process that requires inputs and revisions during the complete project lifecycle. Thus, the CPM must collect inputs and have consultative workshops during the execution phase also which ensures effective coordination among stakeholders

Threshold performance level for design management performance 157

and helps CPM to accelerate the decision-making process. So, if the CPM has identified stakeholders and prepared a stakeholder engagement plan but does not follow the same once the design got finalized during the execution phase, falls under this level.

- **Level 4** – At this level, the CPM is expected to identify all relevant project stakeholders and define their roles early during the design development phase. The stakeholder engagement plan developed by CPM must contain clear communication and coordination from start to finish, and ensure appropriate agencies and technical experts provide consistent input at all stages of project design and development. The CPM should apply a stakeholder engagement plan to manage design changes throughout the project life cycle.
- **Level 5** – Implementation of advanced tools for design development and management like BIM, Navisworks, etc. is expected at this level. The CPM must ensure the simplicity of the user interface and compatibility to effectively use these complex tools by stakeholders involved in the design process. The focus of usage of these tools should be on the coordination and integration of complex information, procedures, and systems and collaboratively devising strategies to manage design changes.

9.2.2 Establishing need centric design process (D_{52})

Design development is a complex process that consists of several stages. Though, the role of CPM is very limited in the design development stage but what is important for CPM is to ensure the execution of work does not get halted due to the unavailability of drawings. Often, CPM faces challenges in managing the delays that happened due to design. In case of delays, the contractors often blame consultants or CPM citing reasons of lack of information in drawings, uncoordinated drawings, non-availability of drawings, delay in drawings, etc. which puts CPM in a difficult situation. Thus the CPM must take the drawings disbursement schedule from the consultant, check whether it is in sync with the project schedule, and share the same with execution agencies to ensure that all the project team members are on the same page. In addition, with the drawings disbursement schedule, the CPM also prepares the responsibility matrix of stakeholders and team members involved in the design process to ensure seamless coordination.

- **Level 1** – At this lowest level, the CPM is not expected to have a formal drawing disbursement schedule. Although in absence of a formal schedule it is difficult for CPM to ensure the availability of required drawing as per work progress; CPM should apply their wisdom, experience, and management skills to ensure that the execution does not gets stopped due to the non-availability of drawings. The skill set of CPM at this level applies to handling small and less complex projects where the number of stakeholders is limited.

158   *Threshold performance level for design management performance*

- **Level 2** – At this level, the CPM should have the drawing disbursement schedule from the design consultant. The CPM should be well aware of the details of drawings required at each interval and expected to regularly follow up with consultants to ensure the availability of the same. However, due to the absence of a detailed project schedule, it is expected that the drawing disbursement schedule is not in sync with the project schedule. But if the CPM has a formal drawing disbursement schedule and ensures the availability of required drawings before the execution of work starts, falls under this level.
- **Level 3** – At this intermediate level, the CPM should ensure the synchronization of the drawing disbursement schedule with the project schedule. The skill set of CPM at this level is valid for small- to medium-scale projects having less complexity. However, incorporating changes in design, and updating the drawing disbursement schedule along with the project schedule, is not expected at this level. But the CPM should ensure that the work progress does not get halted due to the non-availability of required drawings.
- **Level 4** – In addition to the criteria mentioned in Levels 2 and 3, the CPM should have a system to update the drawing disbursement schedule in line with the project schedule after accommodating design changes. Regular consultative meetings with design consultants and execution agencies are expected to maintain the coherence between the design and execution team.
- **Level 5** – Incorporating design changes, and ensuring the availability of coordinated drawing at the site, is the most critical task of CPM. Thus at this highest level, the CPM is expected to apply advanced tools for better coordination and timely delivery of drawings. This level is applicable for large and complex projects containing multiple design and execution agencies.

9.2.3 Establishing decision-making hierarchy (D_{53})

Many times, the project's progress got stuck in between due to indecision by the stakeholders. Thus, the establishment of a decision-making hierarchy is considered one of the major building blocks in any project's success. This is usually done by formalizing reporting relationships. Typically, in construction projects, the strategy and decisions are carried out by different individuals at different timescales based on different kinds of information. Hence, CPM needs to develop a design responsibility matrix defining the roles and reporting relationships of the stakeholders. Accountability is the key to establishing such matrices as often in crisis the individuals keep running away from their responsibility and blaming others accountable for the cause. The approach of CPM for handling the crisis must be proactive where CPM should formulate the strategy well in advance to prevent the crisis.

Threshold performance level for design management performance 159

- **Level 1** – This is applicable for a small and less complex project where a limited number of stakeholders are involved. At this level, the CPM is not expected to develop a formal document containing the list of stakeholders with assigned responsibilities and their engagement throughout the design process. Though, the CPM is expected to take the due approvals during design development from the client and create an informal dialogue between the client, consultant, and execution team.
- **Level 2** – At this level, the CPM should list the number of key stakeholders involved in the design process and develop the engagement plan. Though the responsibilities and roles of stakeholders are not expected to be assigned by the CPM at this level. The formal engagement plan is expected to formalize the decision-making process.
- **Level 3** – In addition to the formal engagement plan, the CPM is expected to assign the responsibilities to the stakeholders and map them with design stages. Simply, the clear role regarding the involvement of stakeholders at each design stage must be clearly defined at this level. However, the formal decision hierarchy matrix of the stakeholders is not expected at this level.
- **Level 4** – At this level, the formal stakeholder's engagement plan with clearly assigned responsibility along with a formal decision hierarchy matrix must be deployed by the CPM. This level of skill set is suggested for a medium to large-scale project with varying complexity. The CPM must ensure to document the information at each design stage and have a process to share it amongst involved stakeholders to strengthen the decision-making process.
- **Level 5** – To strengthen the communication process and information sharing the application of advanced tools is quite effective. Thus, at this highest level, in addition to the criteria mentioned in Level 4, the CPM must implement advanced tools to share real-time information, database creation, and develop communication channels in an integrated manner.

9.2.4 Resolving conflicting interests (D_{54})

Identifying design conflicts and developing a plan of rectification in the early phases of the project is essential for CPM. Typically, in construction projects many activities/works are repetitive and if the conflicts are not identified at an early stage, then they tend to get accumulated throughout the project life cycle and incurs huge rectification cost in the end. As a result, often construction projects experienced issues like cost and time overruns. Due to the involvement of multiple stakeholders, the CPM faces a challenge to recognize the different interests and perspectives of these diverse stakeholders and bring them to the same page for the successful completion of the project. Creating collaborative working opportunities through regular

160 *Threshold performance level for design management performance*

consultative meetings is an effective step that a CPM must implement for managing design conflicts.

- **Level 1** – Due to the involvement of multiple consultants, CPM has to ensure that the drawings received from the consultants are coordinated and without any conflicts. To avoid the stoppage of work and rework, the CPM must ensure that the execution is happening as per the coordinated drawings only.
- **Level 2** – At this level, the CPM should be skilled enough to read and interpret drawings received from the consultants considering the site conditions to identify any conflict. Often the execution team faces the issue of rework due to errors in design, non-compatibility of design concerning site conditions, insufficient information in drawings, etc. Thus, the CPM should have a process in place to verify the drawings received subject to site conditions and communicate with the consultant in case of required revisions or clarifications.
- **Level 3** – In addition to the criteria mentioned in the above two levels, the CPM should intimate the conflicts in drawings/designs to the consultants for required revisions. In case of major revisions resulting in changes in scope must be immediately communicated to the client and associated stakeholders to quantify the impact and decision making.
- **Level 4** – This is the desired level where the CPM is expected to resolve design conflicts using interpersonal skills. Regular consultative meetings between stakeholders and project team members are required to identify early conflicts. Thus, the CPM must ensure that the execution of work is progressing as per coordinated drawings only.
- **Level 5** – In addition to the criteria mentioned in Level 4, the CPM must deploy advanced tools to identify design conflicts. Sharing drawings and documents with other stakeholders and project team members is essential at this level. The knowledge of BIM and related software is essential at this level.

9.2.5 *Effective planning for scope creep (D_{55})*

Naturally, design changes happen to construction projects but CPM needs to have a formal change management plan in place to effectively manage these changes. Anticipating and resolving design changes periodically are essential for any CPM else these changes keep accumulating over different project stages and led to scope creep. It is important to review the designs, identify the conflicts, resolve conflicts through consultative meetings, and inform the client and key stakeholders in case of any deviation from the original scope of work.

- **Level 1** – The CPM has not implemented a formal process to manage changes and control scope lies in this level. Though, the CPM is

Threshold performance level for design management performance 161

expected to review the design regularly to avoid conflicts and should take necessary measures to minimize the possibility of rework due to frequent design changes.

- **Level 2** – At this level, the CPM should review the designs received from the consultants for any conflicts. The CPM is expected to anticipate the design changes and their impact on project timelines and budget. The formal change management processes along with the action plan to control scope creep should be implemented by the CPM.
- **Level 3** – In addition to the criteria mentioned in Level 2, the CPM is expected to establish a formal communication channel to share the information regarding scope deviation with the stakeholders. Often, the CPM struggles to identify the project team member or stakeholders responsible for design changes in such cases the periodic design review meetings and documentation of the outcomes of these meetings would be a very handy tool to avoid conflicts in the later stages of the project.
- **Level 4** – Informing stakeholders regarding design changes periodically may not be enough to control scope creep in the case of large and complex projects. Thus, the CPM should meet with project stakeholders on regular basis and work out collaboratively a viable plan of action to meet the project objectives. In case of major scope deviation due to the change in the client's requirement, changes in specifications, etc. the CPM should take the necessary approvals from all stakeholders involved and prepare a revised action plan in coherence with project team members to meet the desired objectives of the project.
- **Level 5** – At this highest level, in addition to the criteria mentioned in the above levels the CPM should implement advanced tools for real-time monitoring of design conflicts, must anticipate the design conflicts well in advance, and ensure the availability of well-coordinated drawings at the site as per project schedule. Furthermore, the CPM should also implement advanced IT tools to develop robust communication channels to share real-time information with stakeholders and project team members.

9.2.6 Resolving time-cost impact (D_{56})

It is important to establish a strong relationship between design and construction in an integrated manner to create a balance between project time, cost, and quality. Examining various design alternatives, reviewing design concerning their constructability, and sharing inputs from the execution team to the design team are the processes a CPM needs to deploy while planning for both design and construction. Comprehensively, the term "Value Engineering" is more appropriate; CPM should implement it to add value to the designs and eliminate the necessity of extensive revisions during the execution stage. Enhancing quality yet managing project time and cost should be the main objective of the CPM.

Table 32 Threshold performance levels of design performance indicator

S.no	Performance indicators	Weightage of indicators (equal weightage to be moderated as per organizational discretion if any)	Determinants of performance (identified from literature, expert interviews, and field studies; weightages (Wm) derived from questionnaires survey as perception of the field experts)			Performance level (Level m) Levels 1–5 represent performance of a specific Project Manager in a given project situation. As Level 1 being the lowest and Level 5 being the highest having numerical weight varying from 1 to 5. Based on the actions undertaken by the Project Manager performance level is assigned.				
	(Pn)	(Wn)	m	Wm	Determinants	Level 1	Level 2	Level 3	Level 4	Level 5
5	Design performance	W5	1	W51	Establishing stakeholder engagement processes	List of stakeholders involved in design process is in place but stages in design process is not ascertained	Establishing design process stages and list of stakeholders involved	Establishing design process stages and list of stakeholders involved but the formal process of stakeholder engagement during different stages of design process is not in place.	Establishing design process stages and list of stakeholders involved and fix responsibilities of stakeholders involvement with respect to different stages of design process.	Establish design process stages and list of stakeholders involved and fix responsibilities of stakeholders involvement with respect to different stages of design process. Application of advanced tools to create a process of information sharing between all the stakeholders at different design stages to ensure their effective engagement.
			2	W52	Establishing need centric design process	Drawing disbursement schedule is available but not in consensus with the approved project schedule	Drawing disbursement schedule is available and not in consensus with the approved project schedule	Drawing disbursement schedule is available and in consensus with the approved project schedule	Drawing disbursement schedule is available and in consensus with the approved project schedule. Process of accommodating changes in design is in place.	Utilizing advanced tool to prepare the drawing disbursement schedule and in consensus with the approved project schedule

3	W53	Establishing decision making hierarchy	Formal engagement process of stakeholders in the design process is not in place	Engagement of all the stakeholders in the design process. But no responsibilities assigned to all the stakeholders at every stage of the design process.	Engagement of all the stakeholders in the design process. Assigning responsibilities to all the stakeholders at every stage of the design process. But absence of decision hierarchy matrix of the stakeholders is in place.	Engagement of all the stakeholders in the design process. Assigning responsibilities to all the stakeholders at every stage of the design process. Ensuring decision hierarchy matrix of the stakeholders is in place.	Application of advanced tool for the engagement of all the stakeholders in the design process. Assigning responsibilities to all the stakeholders at every stage of the design process. Ensuring decision hierarchy matrix of the stakeholders is in place.
4	W54	Resolving conflicting interests	Identification of conflict in drawings provided by various consultants ensures the availability of coordinated drawing at site.	Identification of conflicts in drawings provided by all the consultant. Verification of execution of drawings as per site conditions.	Identification of conflicts in drawings provided by all the consultant. Verification of execution of drawings as per site conditions. Intimating all conflicts to the consultants for necessary revisions.	Identification of conflicts in drawings provided by all the consultant. Verification of execution of drawings as per site conditions. Resolving all conflicts through consultative meetings and ensuring the availability of coordinated drawings at site as per project schedule.	Application of advanced tool for the identification of conflicts in drawings provided by all the consultant. Verification of execution of drawings as per site conditions. Resolving all conflicts through consultative meetings and ensuring the availability of coordinated drawings at site as per project schedule.

(Continued)

S.no Performance indicators	Weightage of indicators (equal weightage to be moderated as per organizational discretion if any)	Determinants of performance (identified from literature, expert interviews, and field studies; weightages (Wm) derived from questionnaires survey as perception of the field experts)			Performance level (Level m) Levels 1–5 represent performance of a specific Project Manager in a given project situation. As Level 1 being the lowest and Level 5 being the highest having numerical weight varying from 1 to 5. Based on the actions undertaken by the Project Manager performance level is assigned.				
(Pn)	(Wn)	m	Wm	Determinants	Level 1	Level 2	Level 3	Level 4	Level 5
		5	W55	Effective planning for scope creep	Inability to anticipate and plan for scope creep.	Ability to anticipate and plan scope creep.	Anticipate and plan scope creep through design review and intimate the same to stakeholders to avoid the conflict.	Any scope creep is informed and agreeable to all the stakeholders. Taking measures to control all the scope creep due to rework, changes, etc.	Utilizing advanced tool to inform the scope creep and agreeable to all the stakeholders. Taking measures to control all the scope creep due to rework, changes, etc.
		6	W56	Resolving time-cost impacts	Design document (Drawings, specifications, approvals, etc.) is not available at site as per approved project schedule.	Design document is available at site as per approved project schedule. But not ensuring timely approval from associated stakeholders to eliminate the condition of delays.	Design document is available at site as per approved project schedule. Ensuring timely approval from associated stakeholders to eliminate the condition of delays.	Design document is in compliance of project scope. Design document is available at site as per approved project schedule. Ensuring timely approval from associated stakeholders to eliminate the condition of delays. Application of value engineering to avoid the time and cost overrun.	Application of advanced tool to prepare the design document is in compliance of project scope. Design document is available at site as per approved project schedule. Ensuring timely approval from associated stakeholders to eliminate the condition of delays.

9.3 Inferences

The criteria for defining the threshold performance levels (1–5) of the determinants related to design management performance have been detailed out based on which an individual CPM can assess his/her performance against the design management performance indicator and identify their threshold level of performance by evaluating oneself based on the description of each performance level for each determinant of design management performance.

10 Way forward through complexity linkage

10.1 Introduction

The current chapter presents the conclusions and how VDPI can be applied in the process of assessment, leading towards the continuous development of an individual as well as an organization. It helps in overcoming the challenges associated with the non-specific and rather subjective performance assessment of a CPM. A discussion on limitations of VDPI in the current isolated concept is presented which can be addressed by other linkages with project complexity. Therefore, the chapter further provides the way forward by relating the VDPI with the actual degree of project complexities.

10.2 Conclusion and way forward through complexity linked with VDPI

The concept of the VDPI tool provides a quantitative value-based performance index intended for assessing the competency of a CPM. It is derived through the project's performance as assessed over threshold performance levels acceptable at an organizational level or as implemented using national/international performance assessment standards. The VDPI tool can be used as an assessment framework, founded on a mathematical equation-based computation for determining the true qualification of any CPM. Additionally, it helps an organization in selecting the right candidate while hiring for an open position; as per the project necessity and the same individual's performance can be evaluated during all phases of the ongoing project life cycle. It can be assessed through the threshold performance levels, as suitable for defining the role of the project practitioner. These threshold performance levels are for evaluating the performance indicators as established in the VDPI theory (Chapter 3).

The requirement of skillsets for the planned accomplishment of any project is governed by the project requirements and the criticality in defining success parameters of the project. As these indicators are a function of project typology, which can be identified and assessed for examining the performance of the CPM considering delivery of work. The framework discussed

DOI: 10.1201/9781003322771-10

Way forward through complexity linkage 167

in this book used for establishing the VDPI; defines the basic performance indicators that are found to be significant in defining the success of the project by the subject experts in the field of construction management.

One of the key benefits of performance assessment using the VDPI tool is that it is based on a self-assessment approach. It provides an opportunity to introspect and diagnose areas of improvement before an individual is evaluated externally. And, such a self-appraisal then becomes a basis for the organization-level assessment to quickly focus on the deviations on the self-disclosure and the perception by the external evaluator. In any case, VDPI is a systematic process to relate one's own capabilities and the expectations that the project would demand. In absence of such as opportunity, one is blind and the organization is in dark!

The concept of VDPI has a natural pathway to expand into benchmarking to relate performance with external competitors, be that an individual or another organization. The current practice is again non-specific assessment. Through benchmarking performance level of CPM can be gauged realistically by comparing with the best resulting level of performance within or outside the organization. Benchmarking provides an opportunity to establish a baseline of expectations from a prospective CPM on a specific project situation.

VDPI aims to address the primary issue that the construction industry has been facing related to the ambiguity in bringing about accountability in the practice of project management in the construction industry. Considering that there is a dearth of technically competent project managers, VDPI can contribute towards amelioration by improving competence dimension of the limitation.

VDPI is useful in the construction industry for a variety of CPM roles and functions. It operates through levels of performance required or achieved by assessing the delivered work performance competency at different levels. The levels involve: a CPM handling a single project represented as Individual unit (IU) or a CPM handling multiple projects (Project Host), or a CPM handling multiple projects within a vertical of the organization is often termed as a Program Manager (Orgn Unit), or a CPM responsible for projects of different verticals of the organization like railways, airports, highways, mixed-use development, urban infrastructure, interior fit outs, etc. (Multi orgn unit), etc.

However, it is important to identify the limitations of VDPI in its current form which can be identified as:

i   It does not provide any framework for assessing the performance of project sponsors, project consultants, or any other teams involved in a construction project who also consider themselves as project managers engaged on same projects but their overall responsibilities are not exclusively associated with the project execution.

168   *Way forward through complexity linkage*

The applicability of VDPI is associated with organizations specific to construction execution of project management but project manager is a generic position referred to in many organizations. Thus, from VDPI perspective, the subtle nuance needs to be understood.

ii Another limitation of VDPI appears to be that it takes into consideration only time, cost, scope, design, and contract performance only. Though quality and safety are also considered to have significant importance in the case of construction projects, which might require the addition of other performance indicators to the existing framework of VDPI for assessing any CPMs performance. Hence, these performance indicators might have to be expanded based on the requirement of the project and the organizational goals.

The details presented in the book provide ample guidance to include more such parameters as necessary and fine-tune VDPI to suit the specific needs.

The concept presented through VDPI evolves on project management processes adopted for achieving the success of the project. The variables considered are found to be significantly related to the technical skill sets of a CPM. However, the existing body of knowledge of project management performance assessment also emphasizes the project manager's soft skills apart from technical skills and these are as discussed in Section 1.3 of book. Construction industry is a manpower workforce-driven industry that requires leadership, negotiation, and communication skills demanding a well-versed CPM. At construction project sites, it is expected to encounter uncertain conditions which are not planned. A CPM is expected to employ a combination of soft managerial skills, behaviour agility, and handling of interpersonal relations as a part of leadership role in a complex team. Thus, conditions require both hard and soft skills of a CPM.

The value of soft skills is recognized but the VDPI in the current format does not include these soft skills as a part of performance assessment.

While assessing performance of CPM, it is important to establish the level of project performance. In the context of a given project, the performance requirement, most likely, would be different or would need to be articulated differently. This would require the following inputs:

* Project characteristics and emerging complexity
* Construction industry standards

Figure 29 illustrates the process of defining expected performance as a baseline for the CPM to compare with. The VDPI score and performance indicator score values of a given CPM are based on:

* self-assessment
* industrial benchmarking, and
* identified level of project complexity respective to the construction project

Way forward through complexity linkage 169

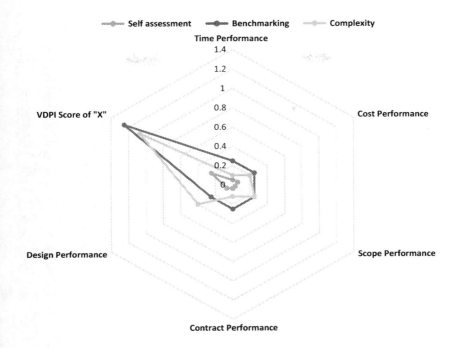

Figure 29 A graphical representation of comparison VDPI score obtained by a CPM "X" through self-assessment, benchmarking, and project complexity requirement.

The illustration in Figure 29 shows VDPI score value in respect of CPM "X" who is a construction project management professional working in the field of the construction industry for more than 10 years with a degree qualification in civil engineering and project management. "X" is also a member of professional bodies.

The VDPI score obtained by CPM "X" is less when compared with the required performance at the level of construction industry standards (marked in the figure hypothetically) which has to be derived through a process of benchmarking and considering project complexity level. Hence, based on the evaluation, there exists a need for improvement of performance of CPM "X". Once the benchmark for comparison is established, CPM "X"'s scope for improvement can be specifically identified.

Authors consider project complexity-based approach for establishing VDPI baseline more objective, though the same can also be interpreted as benchmark. The sections below explain the proposed concept of project complexity indicator (PCI) that has been conceptualized based on the literature already available. The linking of VDPI with the PCI would add to the prowess of VDPI in making a comprehensive assessment of CPM rather user-friendly.

170 Way forward through complexity linkage

10.2.1 Concept of project complexity indicator (PCI)

Baccarini (1996) produced an excellent review of project complexity, which serves as a driver for defining complexity of a project based on the theory of complexity science. The significance of the idea of complexity to the project manager and its function in the strategic management of projects has been emphasized in his research. Baccarini (1996) also states, as a given, reference to (Morris & Hough, 1988), that *"complex projects demand an exceptional level of management, and that the application of conventional systems developed for ordinary projects have been found to be inappropriate for complex projects"*. So, what is needed: is to look into the existing project management methods and judge, whether they are adequate for the successful accomplishment of such complex projects or not.

Project complexity is an ingrained function of project characteristics linked with project success and organizational governing mechanisms, the term project complexity varies with these attributes and is needed to be interlinked with each of its individual qualitative as well as quantitative drivers which vary throughout the project lifecycle.

Complexity is not only associated with the project size and cost; it is also associated with innovation which may demand a lot of toils. The drivers of project complexity are represented in Figure 30.

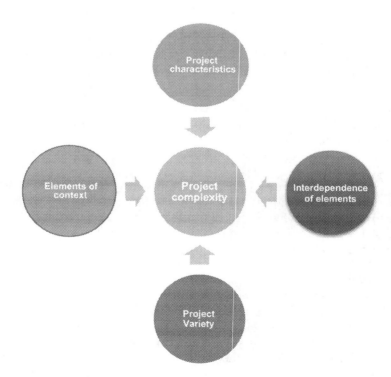

Figure 30 Drivers of project complexity.

Way forward through complexity linkage 171

Since complexity is a new notion that is difficult to define, it can be inferred by taking into account the characteristics and qualities of complicated projects (Lessard et al., 2014). It is derived on the basis of the theory of complexity science which defines *"degree of disorder, instability, emergence, non-linearity, recursiveness, uncertainty, irregularity and randomness"* as the properties related to complexity (Gransberg et al., 2013).

Based on the evidence of available literature, it is quite clear to derive it in terms of project governing drivers, so complexity can be represented as a function of project characteristics, requirements, and organizational mechanisms which requires the establishment of a holistic approach.

Project success being the end deliverable of any project undertaken is defined as a function of major parameters of cost (c), time (t), and quality (q) responsible for major decisions of a project. The other parameters like customer satisfaction cannot be achieved if the complexity associated with the project is not identified.

The project complexity is a function of its drivers and their determinants which might have interrelationships within themselves but for deriving the adequate determinants of complexity the concept of interdependence cannot be ignored.

The interlinking of project complexity determinants and sub-determinants might have different types of cause-effect relationships within themselves, for example, the cost parameter of project success might have a negative relationship with legal complexity. The studies suggest that there are different types of complexities that lead to project complexity.

To ascertain the level of complexity associated with a project, the authors suggest an indicator known as project complexity indicator (PCI) as a suggestive way forward, the formulation of this indicator requires building logical relationships between the identified variables from multiple angles/ perspectives as represented in Figure 30. Complex projects tend to have non-linear feedback loops so, a root cause analysis can be followed to derive the key determinants making a project complex.

The logical relationship between different types of complexity with their sub-determinants defining PCI is needed to be developed using a system thinking approach as it is based on wholes, boundaries, and emergent properties. It requires the establishment of a new project management approach for navigating through complex projects.

Studies suggest that complexity is an important factor leading to project failure. So, the current approach tries to develop a framework for assessing project complexity which can be incorporated into a quantifiable index known as PCI through which we can identify the level of complexity with respect to project characteristics which can be modified based on project diagnosis.

Finally, the PCI would provide an organization with the most precise results for assessing the complexity level of any project which an organization might plan to be bidding for or needs to strategize for formulating its project management approach. Evaluation of the complexity of construction projects would also motivate the decision authorities with adequate

172 *Way forward through complexity linkage*

guidance on how to evaluate, plan, and manage the associated complexity based on the identified PCI levels.

The PCI score value is governed by the determinants of project complexity and can be moderated against their weights and the process of deriving these determinants, sub-determinants to obtain the PCI score is explained in the subsequent sections of this chapter.

10.2.2 Convergence of VDPI and PCI

To truly discover the need for project management competence required for managing any project, it can be defined only if we know about the complexity associated with the project. To define the objective of performance first, it is vital to list down necessary project characteristics governing its successful completion.

The way forward for adapting the VDPI tool in organizations at managerial levels requires clear identification of the degree of complexity associated with the project. Therefore, it can be concluded that the VDPI is a dependent function of project complexity as represented in Figure 31. Thus, there is a need to establish the relation between the degree of project complexity and the threshold level of performance needed in terms of competence of a CPM for defining the success of a project. So, both project complexity and VDPI together contribute towards defining project success.

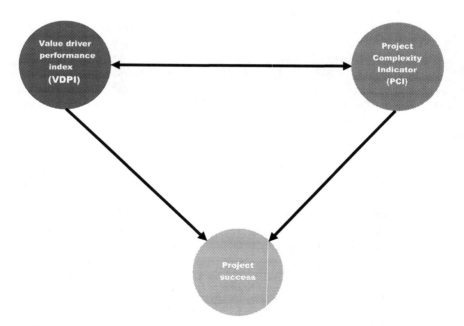

Figure 31 Project success as a function of VDPI and project complexity.

An understanding of the linkage between project complexity and the VDPI concept is essential for developing the ideal equation for defining the level of performance required based on the project complexity level.

Once the PCI score is available, the same can be used in determining the actual level of competence needed regarding any project's level of complexity. It eventually defines the project management approach for achieving successful outcomes of complex projects. It can be identified through an indicative relationship between PCI and VDPI score values.

The mapping of the concept of threshold performance levels of VDPI and PCI levels would result in achieving the success of the project, as represented in Figure 32. Through the mapping of PCI levels against the VDPI, threshold performance levels will have different scenarios related to project performance management which will form the next level of work of this book and will be carried out in phase 2.

So, the overall enhanced application of VDPI would be based on integrating PCI and VDPI value. The convergence of VDPI and PCI provides project management practitioners with the evidence and process-based assessment criterion for analyzing the performance required and delivered by a CPM with due consideration to all possible variables leading to emerging complexities in each phase of the project. It will add as a small step towards standardization of the level of competence accepted for a CPM personnel role in both public and private organizations; promoting project complexity management with adequate performance level identification which addresses the need for performance assessment for the global project management community.

10.2.3 PCI calculation methodology

The concept of PCI builds on the understanding of project context that may be external environment, legal framework, technological advancements, locational disadvantages, etc. The input parameters are encapsulated in the PCI framework as:

$$PCI = \sum f(Oc, Cc, Tc, Lc, Ec, Gc, Uc, Ic, TEc, C1)$$

As observed in the above expression, PCI is a summation function of different types of complexity involved in a project which are a sub-function of project characteristics and emerging properties of associated complexities as represented in Figure 33 where,

- PCI= project complexity indicator
- Oc= organizational complexity
- Cc= cultural complexity
- Tc= Task complexity
- Lc= Legal complexity

174 *Way forward through complexity linkage*

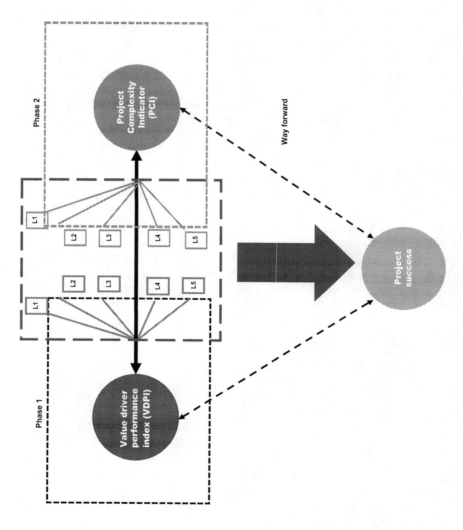

Figure 32 Mapping concept of VDPI and PCI levels.

Way forward through complexity linkage   175

- Ec= Environmental complexity
- Uc= Uncertainty complexity
- Ic= Information complexity
- TEc= Technological complexity
- C_1= Constant

In the above equation, function project complexity is a latent variable that does exist. Its measurable format in a standardized framework-based assessment of complexity is needed to be developed.

The identified taxonomy of project complexity as represented in Figure 33 helps in establishing the level of project complexity. Further, these individual complexity categories have their sub-determinants that are identified through literature review and case study analysis-based approach.

PCI can be described as a summation of the product of determinants of complexity and their weights as represented in Equation 5. This equation is based on the assumption of a linear relationship approach between PCI and all determinants of complexity.

Equation 5

$$PCI = \sum_{n=1}^{N} (Cn * Wn)$$

where C_n are determinants of complexity and W_n represents their respective weightages.

Equation 6

$$PCI = C1*W1 + C2*W2 + C3*W3\ldots\ldots + CN*WN$$

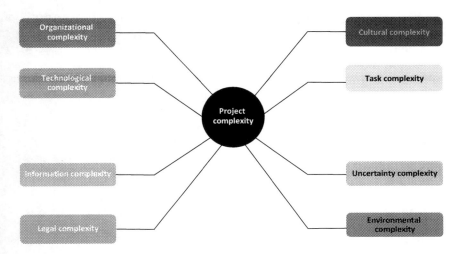

Figure 33 Taxonomy of project complexity.

176 *Way forward through complexity linkage*

In Equation 6 of PCI, C_1, C_2,...C_N are the determinants of project complexity like organizational, technological, uncertainty, environmental complexity, etc. and W_1, W_2, W_3,....,W_N are the importance weightages of determinants of complexity.

The variables for defining C_1, C_2,....,C_N are known as the sub-determinants of respective attributes/determinants of complexity.

Concerning to a specific project, the relative importance of each determinant might vary, and the determinants of complexity may need to evolve and change depending on the techniques adapted by the organization and project typology.

10.2.3.1 Determinants and sub-determinants (variables) of project complexity

The determinants of project complexity are derived using Equation 7 and are expressed as a summation of the product of their sub-determinants and their respective weightages.

Equation 7

$$CN = \sum_{n=0}^{N} C1n * W1n$$

Equation 8

$$C1 = C11 * w11 + C12 * w12 + .. + C1n * w1n$$

For an explanation of the terminologies used in Equation 8, let us assume that C1 is organization complexity and C_{11}, C_{12},...C_{1N} are the sub-determinants of organizational complexity and corresponding to these determinants $w_{11}, w_{12}, .. w_{1n}$ are the weightages of the respective sub-determinants.

Similarly, other determinants of project complexity can be derived using their sub-determinants and their respective weightages.

Equation 9

$$C2 = C21 * w21 + C22 * w22 + ... + C2n * w2n$$

Equation 10

$$C3 = C31 * w31 + C32 * w32 + ... + C3n * w3n$$

Equation 11

$$CN = Cn1 * wn1 + Cn2 * wn2 + ... + Cnn * wnn$$

Way forward through complexity linkage   177

10.2.4 Process of deriving variables for PCI equation

As discussed in Section 10.2.3, project complexity is derived using eight different types of complexity determinants. Further, these determinants are bound to have their sub-determinants that determine these determinants of complexity, so PCI is a function of these determinants of complexity (organizational, technical, informational, uncertainty, legal, environmental, task, and cultural) and their sub-determinants which ultimately depend on the traits and complexity attributes of the project.

The variables for deriving the PCI equation are chosen in such a way that the essence of project management is kept intact while ensuring the complexity of construction projects is duly determined. The approach of gathering these variables should be derived using the simulation modelling considering the following types of analysis when deriving the interdependence between the variables of the PCI equation:

- Predictive analysis (forward reasoning)
- Diagnostic analysis (backward reasoning)
- Sensitivity analysis (explanation reasoning)
- Influence chain analysis (explanation reasoning)

The use of the concept of systems to define dependencies and interdependencies, and then evaluating the true level of project competence based on reliable, accurate, current evidence is necessary to adequately compute project complexity for defining the emerging complexities and their variation during the project lifecycle and its dynamics.

The initial phase of determining these variables which define project complexity is carried out using an extensive literature review. To establish their significance, an evaluation-based approach of finding their significance through case studies as well as literature analysis is followed. As the identified determinant variables and sub-determinant variables are bound to have some mutual interrelationship or collinearity existing among themselves. Their uniqueness, closeness, and subjectivity are established using root cause analysis which gives us the value of their degree of closeness. Once the significance and degree of closeness between the determinants and their sub-determinants are established, a robust holistic list of these determinants would be available to us and further they can be rationalized to have a compact and manageable set matrix for assessing project complexity.

Further, the weightages (W_1, W_2, W_3,...,W_N) of the determinants of complexity can be obtained through a questionnaire survey amongst cohorts with respect to their sub-determinants considering the typology of projects and are validated by the subject matter experts in the field.

Example of type of questions used in survey questionnaire:

Organizational complexity depends upon several sub-determinants like organizational hierarchies, the experience of project participants, and company size. How significant is organizational complexity with respect to the

178 *Way forward through complexity linkage*

typology of projects based on your understanding? Please give it a rating between 1 and 5 as represented in Table 33.

The next step is the calculation of the weightages ($w_{11}, w_{12},..w_{1n}$) of the sub-determinants of these determinants and $C_{11}, C_{12},...C_{1N}$.

These determinants are derived using the following three approaches:

- Root cause analysis
- Case studies evaluation
- Experts
- User perception
- Likert scale survey

To be scored on a scale of 1 to 5 as represented in Figure 39.

Table 33 Questionnaire sample question

Organizational complexity (C1)	1	2	3	4	5
Hospital projects			✓		
Residential projects		✓			
Retail/mixed-use development projects					✓

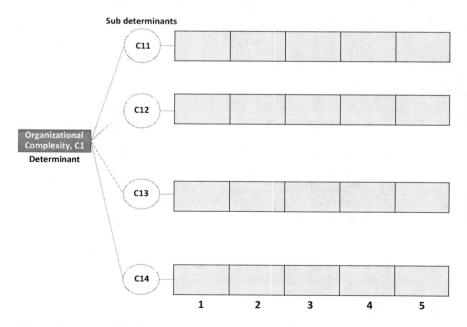

Figure 34 Matrix for determining $C_{11}, C_{12},...,C_{1N}$ sub-determinant values.

Way forward through complexity linkage   179

10.2.5 Level of complexity / PCI range

All projects, in general, are somewhat complicated, but it is crucial to determine the levels of project complexity that will be experienced during the project's lifecycle and the level of project management complexity that will be required.

The range of PCI value scores will be used for determining the level of complexity. The identification of the level of complexity would help in the better formulation of the managerial actions required throughout the lifecycle of the project during the initial stages itself that affect the result of the execution and possibly result in the success of the project.

It is important to identify complexity during the early stages of the project to ensure a successful finish.

The levels of complexity are categorized into five levels:

- Minimum level PCI (Level 1) (which can be easily managed through proper awareness with a basic level of competence) – simple and mildly complex
- Low-level PCI (Level 2)
- Moderate level PCI (Level 3) (with finite and predictable implications requiring an enhanced level of competence for better assessment) – level of moderately complex
- High-level PCI (Level 4) (case of unpredictable implications requiring the best level of competence for management) – level of highly, extremely complex
- Maximum level of PCI (Level 5)

Please note that the degree of complexity and its variation with each project stage must be taken into consideration when establishing the final range because it is relatively usual for project complexity determinants to vary over the different stages of the project lifecycle.

10.3 Inferences

The concept of VDPI for the assessment of the competence-based performance of a CPM through the delivered work output is determined by the mapping of project management processes, skill set/competence of a CPM, and key parameters defining the success of construction projects based on the industry experts specific to Indian construction.

The VDPI assessment frameworks can be used as a document supporting continuous learning and growth with the industry practitioners with slight modifications as per the organizational requirements keeping in mind the following questions:

a   Which tasks are needed to be accomplished as a CPM for the project's success?

180 *Way forward through complexity linkage*

b   What skillsets and levels of performance are required for the accomplishment of the identified tasks/responsibilities?

c   How can the above-identified tasks and skillsets be assessed when assessing the competence of an individual or organization?

Once the above three questions are comprehensively answered and understood by the assessor the framework can be developed and directly applied using the equation of VDPI to derive the quantitative figure of value index based on the performance can be benchmarked and compared with the best standard practices to achieve the best level of performance.

In addition, the concept of VDPI can be developed into a comprehensive index for evaluating performance by the PCI approach based on the determination of the actual complexity level connected with the project and what degree of skill sets are required for the project's successful completion.

References

Baccarini, D. (1996). The concept of project complexity—a review. *International Journal of Project Management*, *14*(4), 201–204. https://doi.org/10.1016/0263-7863(95)00093-3

Gransberg, D. D., Shane, J. S., Strong, K., & Puerto, C. L. del. (2013). Project compelxity mapping in five dimensions of complex transportation projects. *Journal of Management in Engineering*, 29(4), 316–326.

Lessard, D., Sakhrani, V., & Miller, R. (2014). House of project complexity—understanding complexity in large infrastructure projects. *Engineering Project Organization Journal*, *4*(4), 170–192. https://doi.org/10.1080/21573727.2014.907151

Morris P., & Hough G. (1988). *The anatomy of major projects: A study of the reality of project management.* https://www.semanticscholar.org/paper/The-Anatomy-of-Major-Projects%3A-A-Study-of-the-of-Morris-Hough/a47e705661ed5fe3a7eec684e13bc6cfd15104f7

Index

360-degree feedback tool 16, 17, 28

accomplishment 2, 36, 44, 100, 166, 168, 180
actual cost of the work performed (ACWP) 93
actual work 130, 133, 135, 139, 140
administration 105, 146
agency 8, 9, 12, 146, 154
agile project management 2
allocation 1, 35, 99, 100, 121
amelioration 167
anticipate 101, 120, 121, 123, 129, 130, 131, 132, 134, 138, 139, 141, 146, 150, 161, 164
approvals 4, 105, 106, 112, 147, 149, 153, 159, 161, 164
Artificial Intelligence 148
assessment 44

barriers 33, 42
baseline 101, 167, 168, 169
behaviour 21, 29, 32, 168
benchmarking 55, 56, 71, 72, 73, 75, 77, 167, 168, 169
best practices 119
billing performance index (BPI) 93
budget 1, 3, 17, 37, 39, 40, 49, 54, 65, 94, 96, 101, 120, 125, 129, 130, 131, 132, 133, 139, 140, 141, 146, 155, 161
building information modelling 4
bureaucracy 19
business 3, 7, 26, 28, 38, 50, 56, 73, 100, 101, 102, 103

case study 32, 175
cashflow 49
Central Public Works Department 10
change 23, 26, 28, 36, 44, 51, 53, 58, 84, 98, 101, 105, 112, 127, 132, 135, 139, 141, 143, 144, 146, 148, 149, 152, 153, 160, 161, 176
characteristics 3, 4, 9, 22, 168, 170, 171, 172, 173
checklist 104
claim 49, 54, 67, 104, 106, 107, 148, 149, 152, 153
climate change 112
close out 149
cognitive 20, 21, 22
competence 5, 9, 10, 11, 19, 20, 21, 22, 28, 30, 31, 33, 41, 44, 46, 77, 113, 118, 123, 129, 167, 172, 173, 177, 179, 180
complexity 6, 22, 31, 32, 33, 36, 62, 119, 120, 123, 127, 129, 130, 132, 147, 155, 156, 158, 159, 166, 168, 169, 170, 171, 172, 173, 175, 176, 177, 178, 179, 180
conflicting 50, 54, 68, 111, 112, 159, 163
consensus 99, 119, 124, 137, 139, 142, 147, 151, 153, 156, 162
constraints 16, 27, 34, 37, 101, 108, 112, 132
construction 1, 2, 3, 4, 5, 6, 7, 8, 10, 12, 14, 15, 16, 18, 19, 22, 23, 25, 28, 30, 32, 33, 34, 35, 36, 37, 38, 40, 41, 42, 43, 44, 45, 46, 47, 48, 49, 52, 53, 56, 62, 70, 71, 74, 78, 84, 87, 89, 94, 98, 101, 103, 107, 108, 110, 111, 118, 119, 120, 122, 122, 128, 131, 136, 138, 140, 141, 147, 148, 149, 156, 158, 159, 160, 161, 167, 168, 169, 171, 177, 179
construction management 1, 3, 15, 19, 70
consultants 8, 9, 12, 28, 39, 108, 110, 111, 138, 144, 154, 155, 156, 157, 158, 160, 161, 163, 167
contemporary 1, 16
contingencies 49, 54, 65, 94, 96, 131, 132, 134
continuous 2, 7, 14, 16, 25, 33, 36, 52, 69, 72, 77, 117, 120, 140, 166, 179

182 *Index*

contract 3, 23, 25, 31, 39, 49, 51, 54, 60, 62, 67, 70, 84, 98, 100, 102, 103, 104, 105, 106, 107, 112, 119, 123, 145, 146, 147, 148, 149, 150, 153, 154, 168
coordination 6, 33, 49, 54, 64, 89, 91, 103, 108, 110, 118, 119, 124, 129, 131, 132, 135, 137, 142, 144, 156, 157, 158
cost overruns 7, 40, 49, 54, 65, 94, 96, 119, 123, 130, 131, 132, 134, 135, 137, 146
cost performance index (CPI) 93
cost variance (CV) 93, 94
criteria 5, 7, 16, 17, 21, 25, 26, 28, 31, 32, 35, 41, 42, 56, 62, 73, 100, 109, 117, 118, 122, 123, 127, 128, 129, 130, 131, 132, 135, 136, 137, 138, 139, 140, 141, 144, 145, 146, 147, 148, 149, 154, 155, 156, 158, 159, 160, 161, 165
cultural complexity 173

dashboards 74, 131, 135
data collection 132, 135
deadlines 7, 119, 120, 131, 146, 148
deconstruction 112
deliverables 1, 2, 3, 5, 14, 21, 25, 26, 37, 38, 40, 49, 54, 66, 98, 99, 100, 101, 102, 106, 111, 139, 143
demolition 112
deriving 46, 47, 48, 50, 78, 156, 171, 172, 177
design changes 110, 112, 141, 156, 157, 158, 160, 161
design management 155, 165
detailed design 109
determinant score 51
determinants 31, 34, 46, 47, 49, 50, 51, 52, 53, 54, 55, 59, 60, 61, 69, 75, 78, 89, 94, 98, 99, 100, 101, 102, 104, 105, 106, 107, 109, 111, 112, 113, 124, 127, 131, 132, 133, 135, 142, 144, 154, 165, 171, 172, 175, 176, 177, 178, 179
development 1, 4, 7, 9, 16, 17, 18, 19, 20, 21, 22, 24, 25, 26, 29, 32, 33, 34, 40, 41, 42, 43, 44, 45, 47, 52, 55, 72, 73, 77, 78, 89, 110, 117, 120, 131, 149, 156, 157, 159, 166, 167, 178
diagnosis 30, 171
disbursement schedule 157, 158
disciplines 6, 110
dispute 105
document control 147, 151, 152
documentation 27, 39, 100, 101, 111, 141, 147, 148, 153, 161

drawings 111, 112, 155, 157, 158, 160, 161, 163
dynamic 6, 23, 34, 35, 36, 118

Earned Revenue of the Work Performed (ERWP) 93
earned value management 93
efficiency 2, 5, 18, 26, 27, 28, 29, 36, 41, 84, 107, 122
emerging 2, 3, 6, 9, 10, 36, 112, 168, 173, 177
engineering 12, 15, 28, 42, 43, 44, 45, 78, 164
environmental complexity 175
evaluation 5, 7, 17, 18, 25, 26, 29, 33, 34, 36, 39, 42, 43, 44, 46, 48, 51, 52, 53, 55, 56, 74, 75, 77, 101, 107, 108, 109, 118, 128, 136, 145, 155, 169, 177, 178
evolution 14, 97, 110
execution 2, 5, 9, 14, 19, 25, 34, 37, 38, 46, 71, 100, 102, 103, 104, 108, 109, 111, 119, 120, 131, 134, 137, 138, 140, 141, 142, 144, 150, 156, 157, 158, 159, 160, 161, 163, 167, 168, 179
exposure 22, 32, 121
extension of time (EOT) 88
extrapolation 130

facilitator 110
facility management 1
feasibility 39
financial performance 35
framework 15, 19, 20, 25, 32, 33, 35, 36, 41, 42, 43, 44, 103, 166, 167, 168, 171, 173, 175, 180

gauge 25, 33, 56
general mental ability 20

hierarchy 1, 8, 9, 12, 13, 35, 38, 46, 49, 50, 54, 68, 78, 110, 111, 112, 158, 159, 163

inconsistencies 34, 155
indicator 35, 47, 49, 50, 51, 52, 53, 55, 56, 59, 60, 61, 69, 75, 78, 101, 112, 117, 124, 127, 128, 133, 142, 144, 150, 154, 162, 165, 168, 169, 170, 171, 173
individual levels 35, 55, 70
inertia 5
influence chain analysis 177
information complexity 175
innovation 28, 32, 110, 111, 168

Index 183

input unit 70
instability 58, 171
integration 28, 33, 100, 110, 111, 132, 157
interfaces 9, 70, 74, 110, 151, 152
iron triangle 3, 35
irregularity 120, 171
iterative 2, 62, 77, 156

key performance indicators 36, 47, 48, 77, 104

lagging indicators 56, 81
latent variable 175
leadership 4, 7, 22, 28, 29, 39, 40, 41, 42, 43, 70, 110, 168
leading indicators 81
lean 2
legal 103, 104, 105, 106, 145, 171, 173, 177
limitations 16, 46, 166, 167
linear 2, 56, 58, 60, 61, 69, 171, 175
litigation 149

management 1, 2, 3, 4, 5, 6, 7, 9, 10, 12, 14, 15, 16, 18, 19, 20, 22, 23, 25, 26, 27, 29, 30, 33, 35, 36, 37, 38, 39, 40, 41, 42, 43, 44, 45, 47, 48, 49, 50, 51, 52, 53, 54, 56, 60, 62, 65, 67, 70, 78, 81, 84, 86, 89, 92, 93, 94, 96, 97, 98, 99, 100, 101, 102, 103, 104, 105, 106, 107, 108, 109, 110, 111, 112, 113, 117, 118, 121, 124, 126, 127, 128, 129, 132, 133, 135, 136, 141, 142, 144, 145, 147, 148, 149, 153, 154, 155, 156, 157, 160, 161, 165, 167, 168, 169, 170, 171, 172, 173, 177, 179
mapped 50, 99, 104, 112
mediation 103, 149
micromanagement 140
models 1, 14, 17, 28, 29, 35, 43, 61, 78
monitoring 5, 26, 27, 32, 33, 34, 38, 39, 41, 42, 43, 99, 101, 111, 120, 122, 123, 124, 126, 123, 129, 130, 133, 135, 139, 140, 141, 144, 147, 148, 151, 152, 154, 161
multicollinearity 46, 56, 58, 59, 61, 62, 69, 78
multifactor leadership questionnaire 20
multiportfolio 46, 53, 74, 75, 77, 78
mutual interference constant 52
Myers-Briggs type indicator 20

Navisworks 131, 133, 143, 144, 157
need 1, 5, 6, 10, 17, 18, 19, 23, 25, 26, 33, 35, 41, 42, 48, 53, 54, 56, 61, 100, 108, 111, 112, 118, 128, 136, 141, 145, 146, 149, 155, 157, 162, 168, 169, 172, 173, 176
negotiation 22, 28, 105, 149, 168
non-linear 171

obligations 3, 31, 49, 54, 67, 104, 106, 107, 147, 148, 149, 151, 152, 154
opportunity 28, 117, 167
optimize 41
organization unit 55, 72
organizational appraisal 55
organizational complexity 176, 177

parameters 3, 4, 35, 41, 46, 47, 48, 53, 62, 70, 73, 117, 118, 166, 168, 171, 173, 179
payback period 72, 73
peculiarity 109
performance 1, 5, 6, 7, 9, 12, 14, 15, 16, 17, 18, 19, 20, 21, 23, 25, 26, 28, 29, 30, 31, 32, 33, 34, 35, 36, 37, 41, 42, 43, 44, 45, 46, 47, 48, 49, 50, 51, 52, 53, 54, 55, 56, 59, 60, 61, 62, 64, 65, 66, 67, 68, 69, 70, 71, 72, 73, 74, 75, 77, 78, 84, 86, 89, 91, 92, 93, 94, 96, 97, 98, 99, 100, 101, 102, 103, 104, 105, 106, 107, 108, 109, 110, 111, 112, 113, 117, 118, 119, 124, 125, 127, 128, 130, 131, 133, 135, 136, 140, 142, 144, 145, 150, 151, 153, 154, 155, 162, 163, 164, 165, 166, 167, 168, 169, 172, 173, 179, 180
performance index 46
planned 27, 52, 84, 101, 111, 119, 120, 124, 125, 129, 130, 132, 133, 135, 139, 140, 141, 166, 168
Portfolio Management Professional 21
predictive 62, 177
predictor variables 58, 60
prevailing practice 1, 7
probability 59, 121, 132, 137
probity 28
processes 1, 2, 5, 6, 7, 9, 19, 27, 30, 31, 33, 34, 37, 38, 40, 49, 68, 81, 89, 94, 98, 99, 100, 101, 104, 111, 118, 120, 121, 122, 123, 128, 129, 130, 132, 135, 136, 137, 144, 145, 146, 147, 148, 149, 155, 156, 161, 162, 168, 179
processing 34, 46, 52, 70, 71, 72, 73, 74, 75, 77

184 *Index*

procurement 23, 25, 38, 39, 49, 54, 67, 104, 105, 106, 107, 145, 146, 150
productivity 23, 41, 84, 108, 121, 122, 125, 128
project delivery method 8
project management 3, 5, 7, 14, 15, 16, 18, 20, 21, 22, 25, 33, 36, 37, 40, 42, 43, 44, 45, 49, 62, 78, 101, 104, 105, 106
project management professional 21
project participants 177
project typology 9, 42, 48, 155, 166, 176
protocols 33, 110, 146, 147, 151
public sector units 12

qualification 18, 30, 32, 39, 51, 166, 169
quality management 22, 117
quantitative 18, 40, 41, 48, 52, 55, 56, 166, 168, 180

randomness 171
reactive 5, 14
regression 56, 58, 59, 60, 78
regression coefficients 58, 59
relative importance 51, 52, 53, 58, 176
request for Information 111
resource crisis 122
resource management 6, 19
resource planning 49, 64, 89, 91, 121, 122, 125
resource schedule 122
responsibility matrix 111, 138, 157, 158
return on investment 35, 72
review 7, 38, 43, 44, 49, 50, 52, 53, 78, 99, 103, 111, 112, 146, 151, 160, 161, 164, 170, 175, 177
rigorous 6, 147, 148, 152
risk forecasting 49, 64, 89, 91, 120, 121, 125
risk management 54
risk response plan 132, 135
rolling wave planning 98, 101
root cause analysis 171, 177
Royal Institute of Charted Surveyors 18

schedule 3, 26, 27, 49, 54, 64, 89, 91, 101, 105, 109, 111, 119, 120, 121, 122, 123, 124, 125, 126, 129, 130, 131, 132, 133, 135, 137, 143, 145, 146, 150, 157, 158, 161, 162, 163, 164
schedule performance index (SPI) 88
schedule variance (SV) 88
scientific tracking 123

scope of work 3, 9, 18, 38, 71, 98, 100, 101, 106, 111, 136, 141, 154, 160
self-appraisal 55, 167
self-assessment 25, 28, 53, 56, 70, 73, 75, 77, 117, 118, 167, 168, 169
self-disclosure 167
sensitivity 177
simulation modelling 177
size 3, 4, 6, 9, 22, 36, 38, 52, 59, 61, 71, 154, 168, 177
skill 7, 19, 30, 31, 32, 33, 36, 42, 50, 122, 129, 130, 132, 138, 156, 157, 158, 159, 168, 179, 180
skilled 4, 6, 14, 160
snagging 154
software 46, 58, 70, 71, 160
specification 41, 146
speed of construction 89
stakeholder satisfaction 110
standards 5, 10, 16, 17, 18, 20, 25, 27, 28, 32, 33, 34, 40, 43, 46, 55, 77, 105, 111, 166, 168, 169
statistical 56, 58, 59
structural equation modelling 61
sub-function 173
supply chain 110, 111
symbiotic 30

tangible 46, 56, 101
task complexity 173
taxonomy 7, 16, 17, 175
technically equipped 6
technological complexity 175
tender 39, 109
time 17, 26, 37, 45, 54, 60, 64, 81, 84, 89, 91, 104, 120, 124
time management 117, 127
training 18, 84

uncertainty complexity 175
unified 110
unique 3, 6, 23, 25, 33, 36, 40, 52, 56
unit cost 94
usability 109
user 17, 41, 46, 53, 56, 59, 70, 111, 117, 139, 157, 169
user perception 178

value 3, 7, 16, 17, 35, 40, 41, 42, 43, 45, 46, 53, 56, 58, 59, 60, 72, 73, 93, 112, 161, 164, 166, 168, 169, 172, 173, 177, 179, 180
value engineering 41, 112, 161

Index 185

variance 27, 49, 54, 65, 94, 96, 105, 123, 130, 132, 133, 143
VDPI 9, 12, 19, 30, 32, 33, 46, 47, 48, 49, 50, 51, 52, 53, 55, 56, 57, 59, 60, 61, 62, 69, 70, 71, 72, 73, 74, 75, 76, 77, 78, 113, 117, 118, 166, 167, 168, 169, 172, 173, 174, 179, 180
verification 154

waterfall project management 2
way ahead 35

weightages 47, 49, 50, 53, 59, 61, 64, 65, 66, 67, 68, 124, 125, 133, 142, 150, 151, 153, 162, 163, 164, 175, 176, 177, 178
wisdom 28, 62, 99, 108, 119, 120, 121, 122, 130, 131, 132, 140, 146, 157
work breakdown structure (WBS) 26, 97, 99
work packages 38, 99, 119, 122, 124, 129, 133, 139, 140, 150, 153, 154

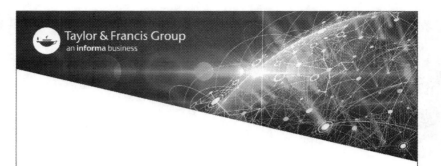

Printed in the United States
by Baker & Taylor Publisher Services